DATE ______________

CONCEPT OR SKILL

____

WHAT JOB USES THIS MATH?

____

ASK TEACHER WHERE IN THE NEIGHBORHOOD, HOME, MONEY, NATURE, PAST/FUTURE EVENTS, AND/OR, ARTS/MUSIC/SPORTS WILL YOU SEE OR USE THIS SKILL?

REAL LIFE CONNECTION - ASK TEACHER WHERE WILL SEE THIS SKILL AT...

1. NEIGHBORHOOD	2. HOME	3. MONEY
4. NATURE	5. PAST/FUTURE EVENTS	6. ART/MUSIC/SPORTS

HOW ARE THE PROBLEMS SOLVED

EXAMPLE 1	EXAMPLE 2
EXAMPLE 3	EXAMPLE 4

TWO WORDS FROM THE LESSON TO CONTRAST/COMPARE

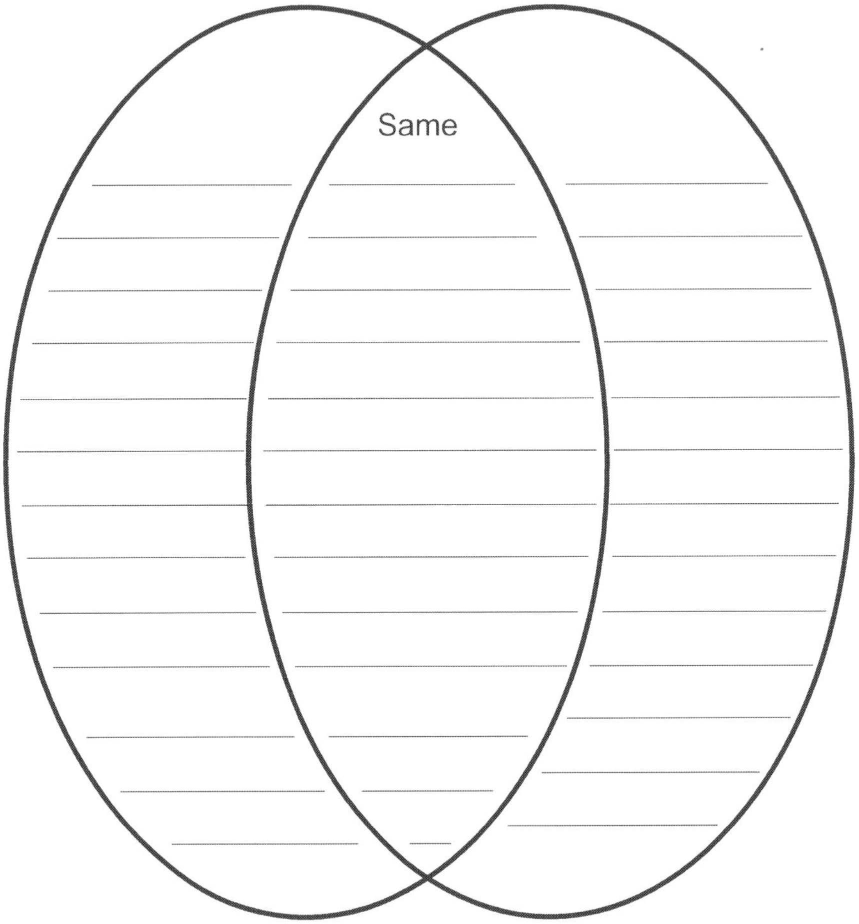

STEPS TAUGHT IN THIS

WORD

1 ______

2 ______

3 ______

MATH WORK OR SYNONYM/ ANTONYM

1 ______

2 ______

3 ______

WORD	DETAIL
1 ______	1 ______
2 ______	2 ______
3 ______	3 ______
4 ______	4 ______
5 ______	5 ______
6 ______	6 ______
7 ______	7 ______
8 ______	8 ______
9 ______	9 ______
10 ______	10 ______

DEFINITION OR KEY WORDS

PICTURE OR EXAMPLE

CONCEPT OR SKILL

STRATEGIES

COMMON MISTAKES OR MISCONCEPTIONS

HOW DOES THIS RELATE TO OTHER MATH TOPICS?

______________________ ______________________

______________________ ______________________

Math Action Terms/Words

Formulas	Equations

EXIT TICKET: THREE THINGS I LEARNED/WHAT STOOD OUT

1
2
3

NEED PRACTICE	ALMOST GOT IT	GOOD AT THIS SKILL

TEACHER NOTE. PARENT NOTE

DATE ____________

CONCEPT OR SKILL

____

WHAT JOB USES THIS MATH?

____

ASK TEACHER WHERE IN THE NEIGHBORHOOD, HOME, MONEY, NATURE, PAST/FUTURE EVENTS, AND/OR, ARTS/MUSIC/SPORTS WILL YOU SEE OR USE THIS SKILL?

REAL LIFE CONNECTION—ASK TEACHER WHERE WILL SEE THIS SKILL IN...

1. NEIGHBORHOOD	2. HOME	3. MONEY
4. NATURE	5. PAST/FUTURE EVENTS	6. ART/MUSIC/SPORTS

HOW ARE THE PROBLEMS SOLVED

EXAMPLE 1	EXAMPLE 2
EXAMPLE 3	EXAMPLE 4

TWO WORDS FROM THE LESSON TO CONTRAST/COMPARE

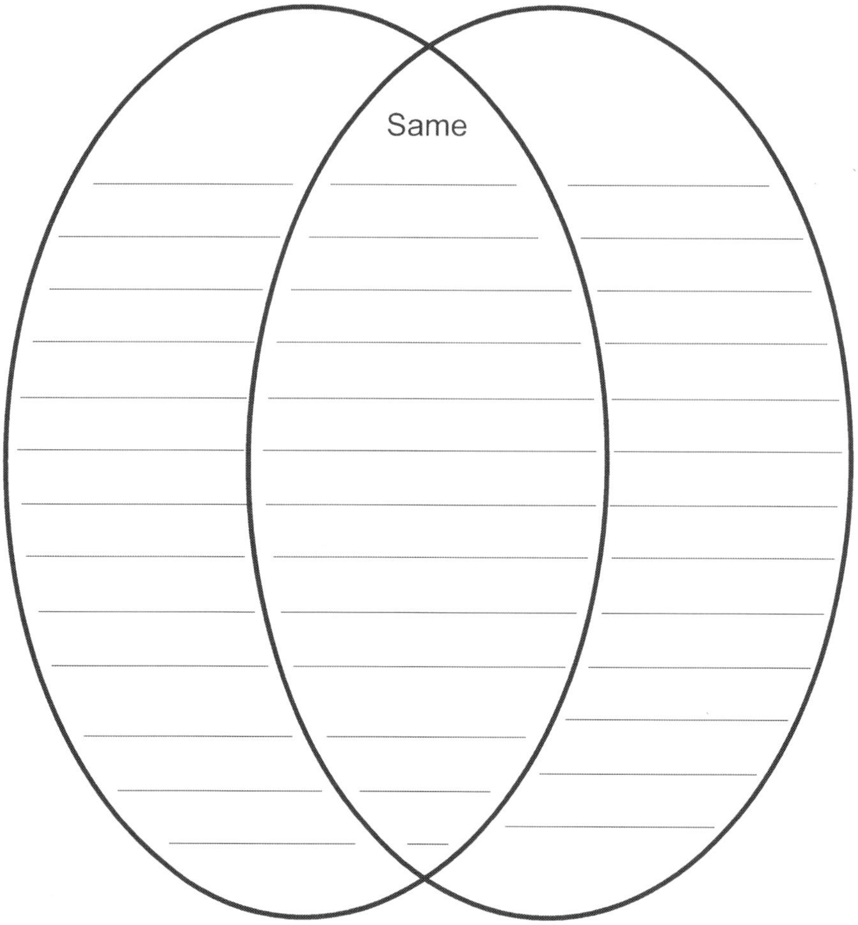

STEPS TAUGHT IN THIS

WORD

1 ____________

2 ____________

3 ____________

MATH WORK OR SYNONYM/ ANTONYM

1 ____________

2 ____________

3 ____________

WORD	DETAIL
1 ____________	1 ____________
2 ____________	2 ____________
3 ____________	3 ____________
4 ____________	4 ____________
5 ____________	5 ____________
6 ____________	6 ____________
7 ____________	7 ____________
8 ____________	8 ____________
9 ____________	9 ____________
10 ____________	10 ____________

DEFINITION OR KEY WORDS

PICTURE OR EXAMPLE

CONCEPT OR SKILL

STRATEGIES

COMMON MISTAKES OR MISCONCEPTIONS

HOW DOES THIS RELATE TO OTHER MATH TOPICS?

Math Action Terms/Words

Formulas	Equations

EXIT TICKET: THREE THINGS I LEARNED/WHAT STOOD OUT

1
2
3

NEED PRACTICE	ALMOST GOT IT	GOOD AT THIS SKILL

TEACHER NOTE.

PARENT NOTE

DATE ____________

CONCEPT OR SKILL

____

WHAT JOB USES THIS MATH?

____

ASK TEACHER WHERE IN THE NEIGHBORHOOD, HOME, MONEY, NATURE, PAST/FUTURE EVENTS, AND/OR, ARTS/MUSIC/SPORTS WILL YOU SEE OR USE THIS SKILL?

REAL LIFE CONNECTION - ASK TEACHER WHERE WILL SEE THIS SKILL IN...

1. NEIGHBORHOOD	2. HOME	3. MONEY
4. NATURE	5. PAST/FUTURE EVENTS	6. ART/MUSIC/SPORTS

HOW ARE THE PROBLEMS SOLVED

EXAMPLE 1	EXAMPLE 2
EXAMPLE 3	EXAMPLE 4

TWO WORDS FROM THE LESSON TO CONTRAST/COMPARE

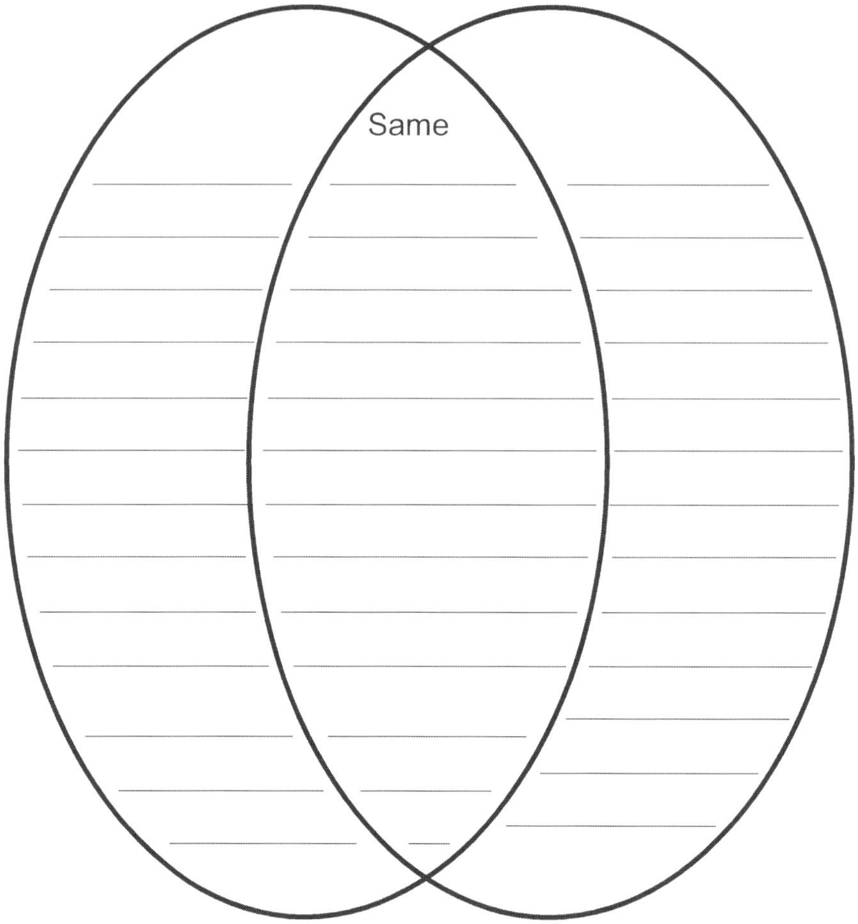

STEPS TAUGHT IN THIS LESSON

1 ______

2 ______

3 ______

MATH WORK OR SYNONYM/ ANTONYM

1 ______

2 ______

3 ______

WORD	DETAIL
1 ______	1 ______
2 ______	2 ______
3 ______	3 ______
4 ______	4 ______
5 ______	5 ______
6 ______	6 ______
7 ______	7 ______
8 ______	8 ______
9 ______	9 ______
10 ______	10 ______

DEFINITION OR KEY WORDS

PICTURE OR EXAMPLE

CONCEPT OR SKILL

STRATEGIES

COMMON MISTAKES OR MISCONCEPTIONS

HOW DOES THIS RELATE TO OTHER MATH TOPICS?

______________________ ______________________

______________________ ______________________

Math Action Terms/Words

Formulas	Equations

EXIT TICKET: THREE THINGS I LEARNED/WHAT STOOD OUT

1	
2	
3	

NEED PRACTICE	ALMOST GOT IT	GOOD AT THIS SKILL

TEACHER NOTE. PARENT NOTE

DATE ____________

CONCEPT OR SKILL

____

WHAT JOB USES THIS MATH?

____

ASK TEACHER WHERE IN THE NEIGHBORHOOD, HOME, MONEY, NATURE, PAST/FUTURE EVENTS, AND/OR, ARTS/MUSIC/SPORTS WILL YOU SEE OR USE THIS SKILL?

REAL LIFE CONNECTION—ASK TEACHER WHERE YOU WILL SEE THIS SKILL IN...

1. NEIGHBORHOOD	2. HOME	3. MONEY
4. NATURE	5. PAST/FUTURE EVENTS	6. ART/MUSIC/SPORTS

HOW ARE THE PROBLEMS SOLVED

EXAMPLE 1	EXAMPLE 2
EXAMPLE 3	EXAMPLE 4

TWO WORDS FROM THE LESSON TO CONTRAST/COMPARE

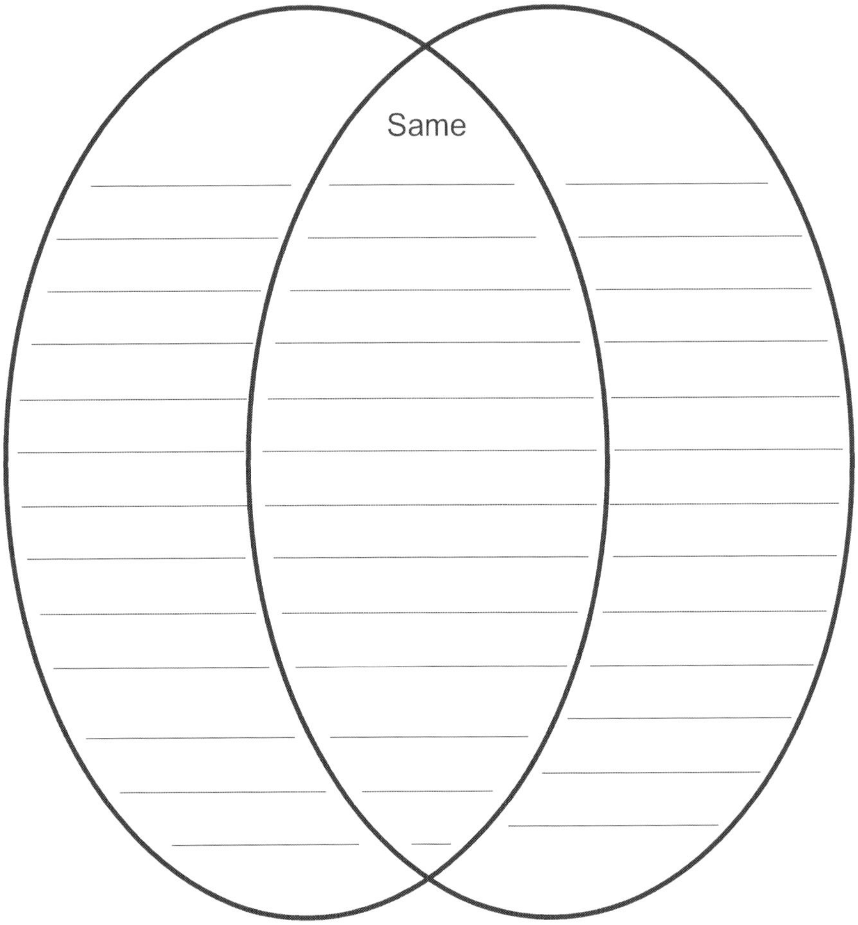

STEPS TAUGHT IN THIS

LESSON

1 ____________________

2 ____________________

3 ____________________

MATH WORK OR SYNONYM/ ANTONYM

1 ____________________

2 ____________________

3 ____________________

WORD	DETAIL
1 ____________	1 ____________
2 ____________	2 ____________
3 ____________	3 ____________
4 ____________	4 ____________
5 ____________	5 ____________
6 ____________	6 ____________
7 ____________	7 ____________
8 ____________	8 ____________
9 ____________	9 ____________
10 ____________	10 ____________

DEFINITION OR KEY WORDS

PICTURE OR EXAMPLE

CONCEPT OR SKILL

STRATEGIES

COMMON MISTAKES OR MISCONCEPTIONS

HOW DOES THIS RELATE TO OTHER MATH TOPICS?

Math Action Terms/Words

Formulas	Equations

EXIT TICKET: THREE THINGS I LEARNED/WHAT STOOD OUT

1	
2	
3	

NEED PRACTICE	ALMOST GOT IT	GOOD AT THIS SKILL

TEACHER NOTE.	PARENT NOTE

DATE ____________

CONCEPT OR SKILL

____

WHAT JOB USES THIS MATH?

____

ASK TEACHER WHERE IN THE NEIGHBORHOOD, HOME, MONEY, NATURE, PAST/FUTURE EVENTS, AND/OR, ARTS/MUSIC/SPORTS WILL YOU SEE OR USE THIS SKILL?

1. NEIGHBORHOOD	2. HOME	3. MONEY
4. NATURE	5. PAST/FUTURE EVENTS	6. ART/MUSIC/SPORTS

HOW ARE THE PROBLEMS SOLVED

EXAMPLE 1	EXAMPLE 2
EXAMPLE 3	EXAMPLE 4

TWO WORDS FROM THE LESSON TO CONTRAST/COMPARE

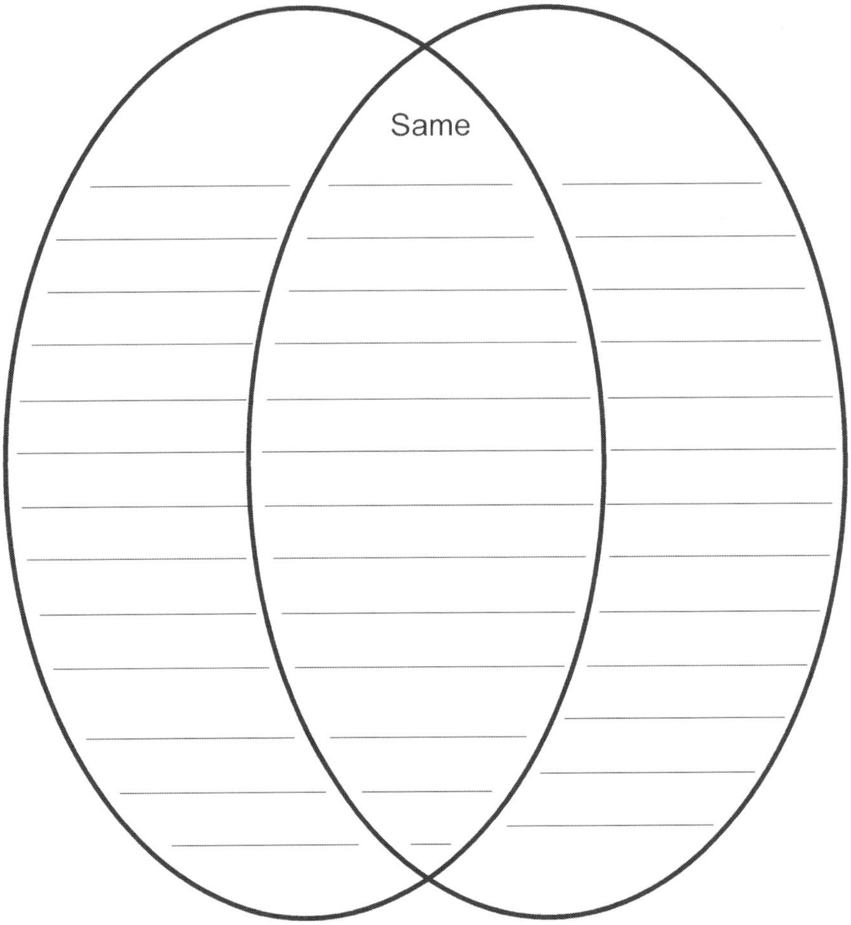

STEPS TAUGHT IN THIS

LESSON

1 ____________________

2 ____________________

3 ____________________

MATH WORK OR SYNONYM/ ANTONYM

1 ____________________

2 ____________________

3 ____________________

WORD	DETAIL
1	1
2	2
3	3
4	4
5	5
6	6
7	7
8	8
9	9
10	10

DEFINITION OR KEY WORDS

PICTURE OR EXAMPLE

CONCEPT OR SKILL

STRATEGIES

COMMON MISTAKES OR MISCONCEPTIONS

HOW DOES THIS RELATE TO OTHER MATH TOPICS?

Math Action Terms/Words

Formulas	Equations

EXIT TICKET: THREE THINGS I LEARNED/WHAT STOOD OUT

1	
2	
3	

NEED PRACTICE	ALMOST GOT IT	GOOD AT THIS SKILL

TEACHER NOTE.

PARENT NOTE

DATE ____________

CONCEPT OR SKILL

____

WHAT JOB USES THIS MATH?

____

ASK TEACHER WHERE IN THE NEIGHBORHOOD, HOME, MONEY, NATURE, PAST/FUTURE EVENTS, AND/OR, ARTS/MUSIC/SPORTS WILL YOU SEE OR USE THIS SKILL?

REAL LIFE CONNECTION - ASK TEACHER WHERE WILL SEE THIS SKILL IN:

1. NEIGHBORHOOD	2. HOME	3. MONEY
4. NATURE	5. PAST/FUTURE EVENTS	6. ART/MUSIC/SPORTS

HOW ARE THE PROBLEMS SOLVED

EXAMPLE 1	EXAMPLE 2
EXAMPLE 3	EXAMPLE 4

TWO WORDS FROM THE LESSON TO CONTRAST/COMPARE

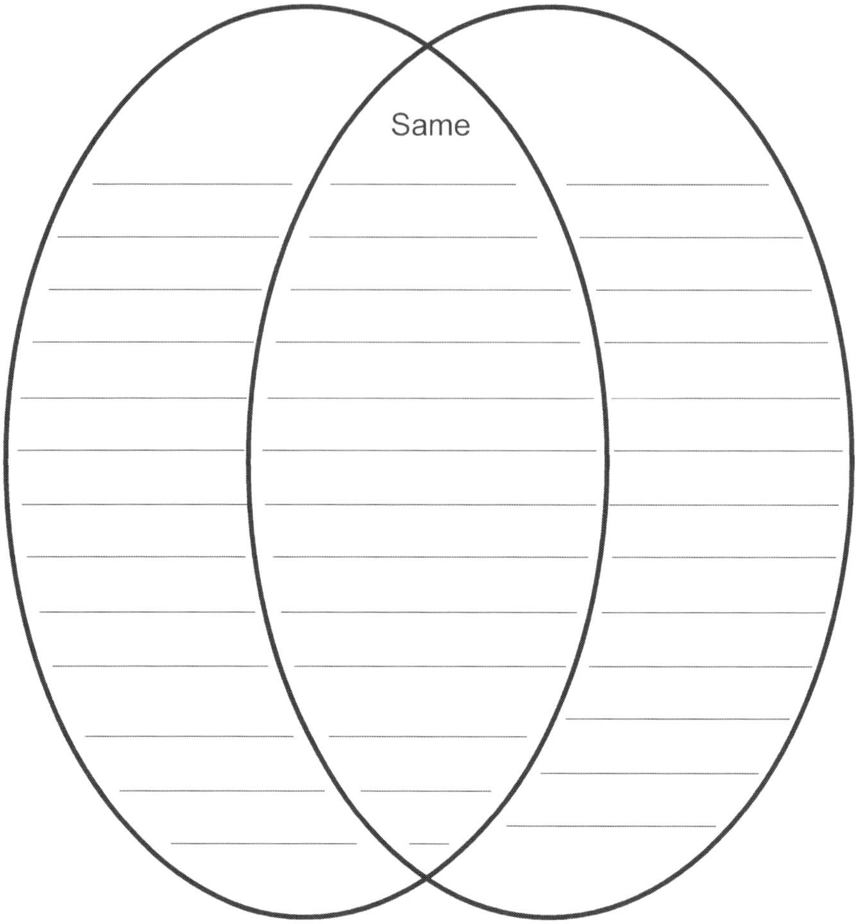

STEPS TAUGHT IN THIS

LESSON

1 ______

2 ______

3 ______

MATH WORK OR SYNONYM/ ANTONYM

1 ______

2 ______

3 ______

WORD	DETAIL
1 ______	1 ______
2 ______	2 ______
3 ______	3 ______
4 ______	4 ______
5 ______	5 ______
6 ______	6 ______
7 ______	7 ______
8 ______	8 ______
9 ______	9 ______
10 ______	10 ______

DEFINITION OR KEY WORDS

PICTURE OR EXAMPLE

CONCEPT OR SKILL

STRATEGIES

COMMON MISTAKES OR MISCONCEPTIONS

HOW DOES THIS RELATE TO OTHER MATH TOPICS?

____________________ ____________________

____________________ ____________________

Math Action Terms/Words

Formulas	Equations

EXIT TICKET: THREE THINGS I LEARNED/WHAT STOOD OUT

1	
2	
3	

NEED PRACTICE	ALMOST GOT IT	GOOD AT THIS SKILL

TEACHER NOTE. PARENT NOTE

DATE ______________

CONCEPT OR SKILL

____

WHAT JOB USES THIS MATH?

____

ASK TEACHER WHERE IN THE NEIGHBORHOOD, HOME, MONEY, NATURE, PAST/FUTURE EVENTS, AND/OR, ARTS/MUSIC/SPORTS WILL YOU SEE OR USE THIS SKILL?

REAL LIFE CONNECTION - ASK TEACHER WHERE WILL SEE THIS SKILL IN...

1. NEIGHBORHOOD	2. HOME	3. MONEY
4. NATURE	5. PAST/FUTURE EVENTS	6. ART/MUSIC/SPORTS

HOW ARE THE PROBLEMS SOLVED

EXAMPLE 1	EXAMPLE 2
EXAMPLE 3	EXAMPLE 4

TWO WORDS FROM THE LESSON TO

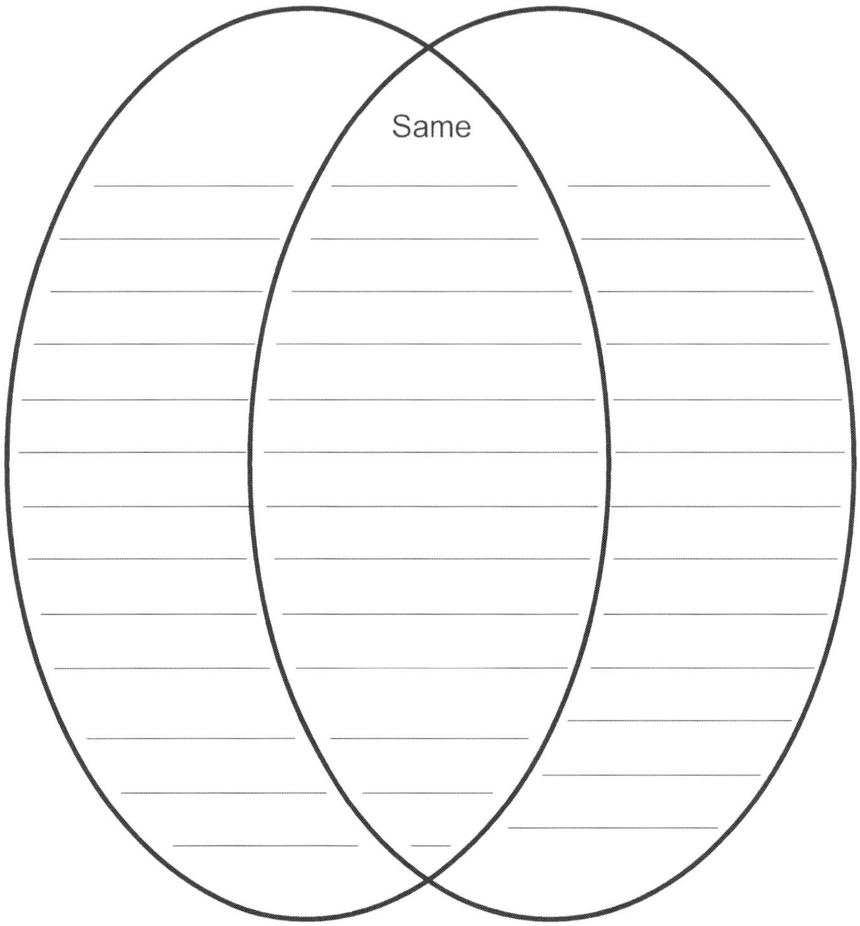

STEPS TAUGHT IN THIS

LESSON

1 ______

2 ______

3 ______

MATH WORK OR SYNONYM/ ANTONYM

1 ______

2 ______

3 ______

WORD	DETAIL
1 ______	1 ______
2 ______	2 ______
3 ______	3 ______
4 ______	4 ______
5 ______	5 ______
6 ______	6 ______
7 ______	7 ______
8 ______	8 ______
9 ______	9 ______
10 ______	10 ______

DEFINITION OR KEY WORDS

PICTURE OR EXAMPLE

CONCEPT OR SKILL

STRATEGIES

COMMON MISTAKES OR MISCONCEPTIONS

HOW DOES THIS RELATE TO OTHER MATH TOPICS?

Math Action Terms/Words

Formulas	Equations

EXIT TICKET: THREE THINGS I LEARNED/WHAT STOOD OUT

1	
2	
3	

NEED PRACTICE	ALMOST GOT IT	GOOD AT THIS SKILL

TEACHER NOTE.	PARENT NOTE

DATE ____________

CONCEPT OR SKILL

____

WHAT JOB USES THIS MATH?

____

ASK TEACHER WHERE IN THE NEIGHBORHOOD, HOME, MONEY, NATURE, PAST/FUTURE EVENTS, AND/OR, ARTS/MUSIC/SPORTS WILL YOU SEE OR USE THIS SKILL?

REAL LIFE CONNECTION—ASK TEACHER WHERE WILL SEE THIS SKILL IN...

1. NEIGHBORHOOD	2. HOME	3. MONEY
4. NATURE	5. PAST/FUTURE EVENTS	6. ART/MUSIC/SPORTS

HOW ARE THE PROBLEMS SOLVED

EXAMPLE 1	EXAMPLE 2
EXAMPLE 3	EXAMPLE 4

TWO WORDS FROM THE LESSON TO CONTRAST/COMPARE

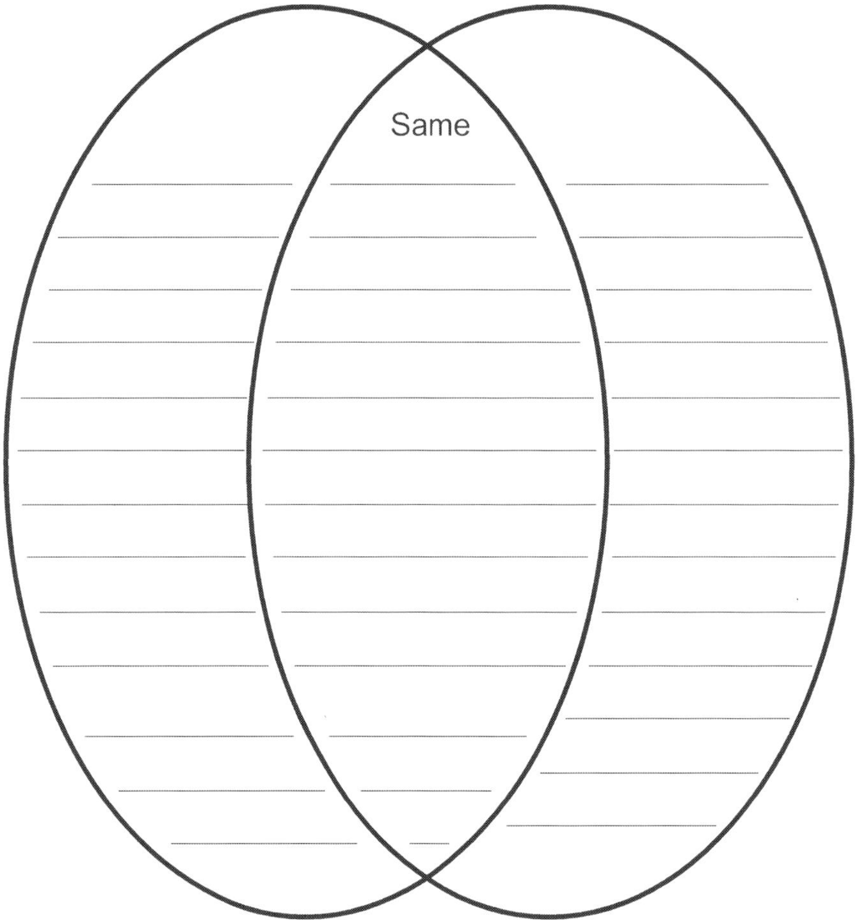

STEPS TAUGHT IN THIS

LESSON

1 ____________________

2 ____________________

3 ____________________

MATH WORK OR SYNONYM/ ANTONYM

1 ____________________

2 ____________________

3 ____________________

WORD	DETAIL
1	1
2	2
3	3
4	4
5	5
6	6
7	7
8	8
9	9
10	10

DEFINITION OR KEY WORDS

PICTURE OR EXAMPLE

CONCEPT OR SKILL

STRATEGIES

COMMON MISTAKES OR MISCONCEPTIONS

HOW DOES THIS RELATE TO OTHER MATH TOPICS?

Math Action Terms/Words

Formulas	Equations

EXIT TICKET: THREE THINGS I LEARNED/WHAT STOOD OUT

1	
2	
3	

NEED PRACTICE	ALMOST GOT IT	GOOD AT THIS SKILL

TEACHER NOTE.	PARENT NOTE

DATE ____________

CONCEPT OR SKILL

____

WHAT JOB USES THIS MATH?

____

ASK TEACHER WHERE IN THE NEIGHBORHOOD, HOME, MONEY, NATURE, PAST/FUTURE EVENTS, AND/OR, ARTS/MUSIC/SPORTS WILL YOU SEE OR USE THIS SKILL?

1. NEIGHBORHOOD	2. HOME	3. MONEY
4. NATURE	5. PAST/FUTURE EVENTS	6. ART/MUSIC/SPORTS

HOW ARE THE PROBLEMS SOLVED

EXAMPLE 1	EXAMPLE 2
EXAMPLE 3	EXAMPLE 4

TWO WORDS FROM THE LESSON TO CONTRAST/COMPARE

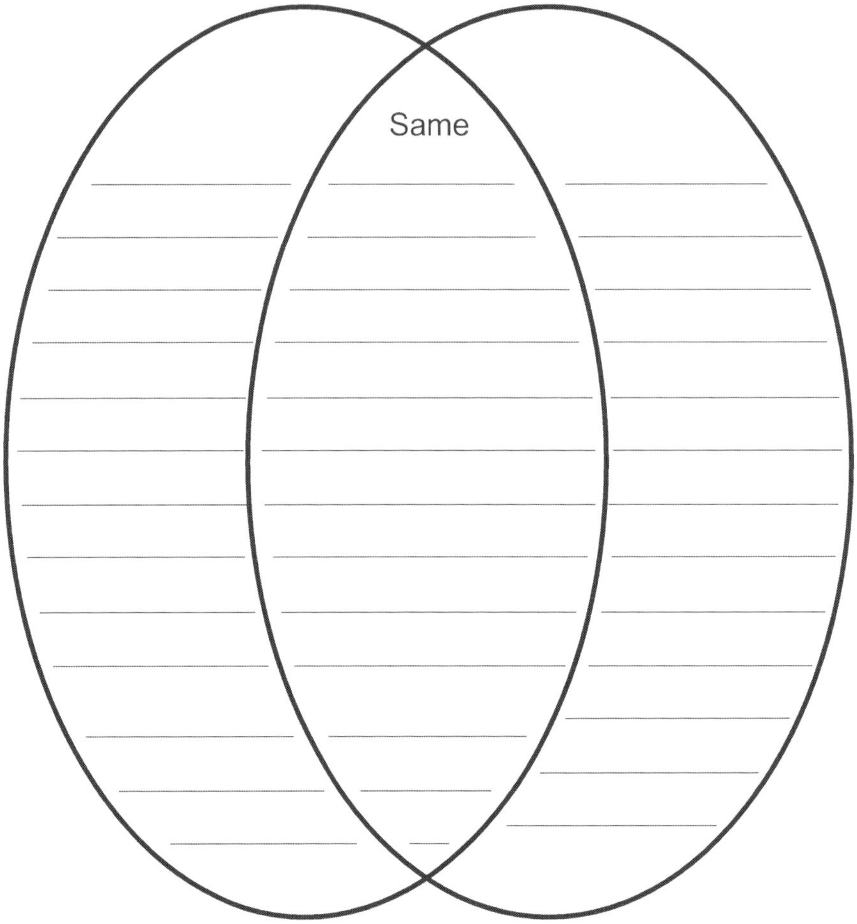

STEPS TAUGHT IN THIS LESSON

1 ________________

2 ________________

3 ________________

MATH WORK OR SYNONYM/ ANTONYM

1 ________________

2 ________________

3 ________________

WORD	DETAIL
1	1
2	2
3	3
4	4
5	5
6	6
7	7
8	8
9	9
10	10

DEFINITION OR KEY WORDS

PICTURE OR EXAMPLE

CONCEPT OR SKILL

STRATEGIES

COMMON MISTAKES OR MISCONCEPTIONS

HOW DOES THIS RELATE TO OTHER MATH TOPICS?

______________________ ______________________

______________________ ______________________

Math Action Terms/Words

Formulas	Equations

EXIT TICKET: THREE THINGS I LEARNED/WHAT STOOD OUT

1	
2	
3	

NEED PRACTICE	ALMOST GOT IT	GOOD AT THIS SKILL

TEACHER NOTE.	PARENT NOTE

DATE ____________

CONCEPT OR SKILL

____

WHAT JOB USES THIS MATH?

____

ASK TEACHER WHERE IN THE NEIGHBORHOOD, HOME, MONEY, NATURE, PAST/FUTURE EVENTS, AND/OR, ARTS/MUSIC/SPORTS WILL YOU SEE OR USE THIS SKILL?

REAL LIFE CONNECTION—ASK TEACHER WHERE WILL SEE THIS SKILL IN...

1. NEIGHBORHOOD	2. HOME	3. MONEY
4. NATURE	5. PAST/FUTURE EVENTS	6. ART/MUSIC/SPORTS

HOW ARE THE PROBLEMS SOLVED

EXAMPLE 1	EXAMPLE 2
EXAMPLE 3	EXAMPLE 4

TWO WORDS FROM THE LESSON TO CONTRAST/COMPARE

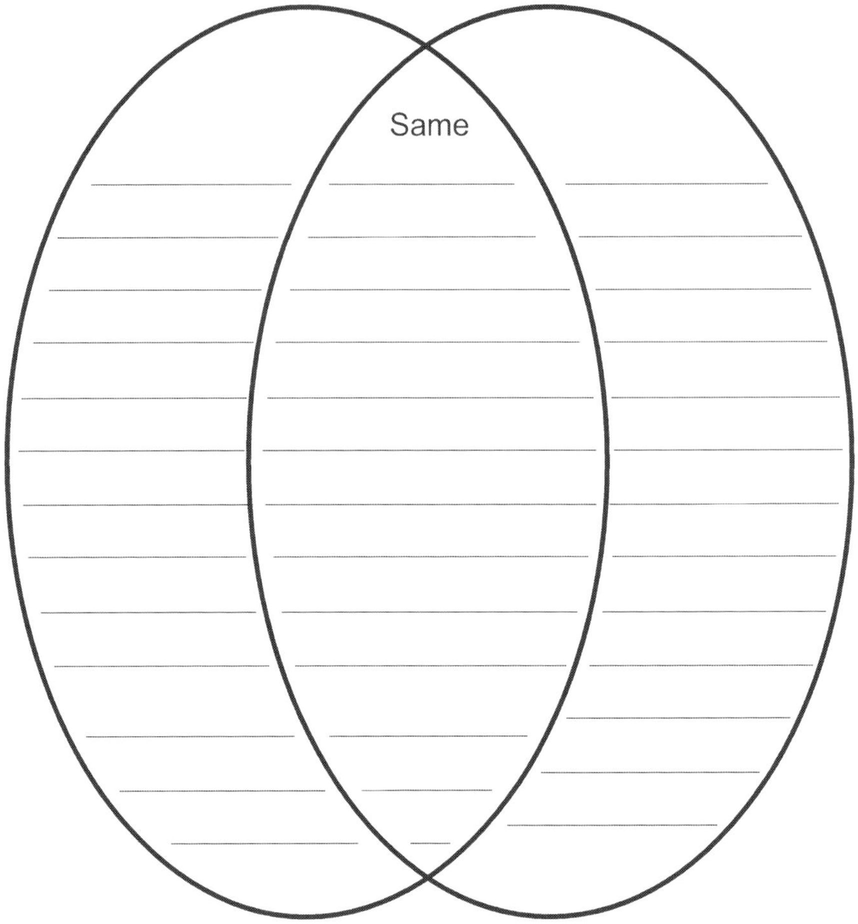

STEPS TAUGHT IN THIS

LESSON

1 ____________________

2 ____________________

3 ____________________

MATH WORK OR SYNONYM/ ANTONYM

1 ____________________

2 ____________________

3 ____________________

WORD	DETAIL
1	1
2	2
3	3
4	4
5	5
6	6
7	7
8	8
9	9
10	10

DEFINITION OR KEY WORDS	PICTURE OR EXAMPLE

CONCEPT OR SKILL

STRATEGIES	COMMON MISTAKES OR MISCONCEPTIONS

HOW DOES THIS RELATE TO OTHER MATH TOPICS?

______________________ ______________________

______________________ ______________________

Math Action Terms/Words

Formulas	Equations

EXIT TICKET: THREE THINGS I LEARNED/WHAT STOOD OUT

1	
2	
3	

NEED PRACTICE	ALMOST GOT IT	GOOD AT THIS SKILL

TEACHER NOTE.	PARENT NOTE

DATE ____________

CONCEPT OR SKILL

____

WHAT JOB USES THIS MATH?

____

ASK TEACHER WHERE IN THE NEIGHBORHOOD, HOME, MONEY, NATURE, PAST/FUTURE EVENTS, AND/OR, ARTS/MUSIC/SPORTS WILL YOU SEE OR USE THIS SKILL?

REAL LIFE CONNECTION- ASK TEACHER WHERE WILL SEE THIS SKILL IN:

1. NEIGHBORHOOD	2. HOME	3. MONEY
4. NATURE	5. PAST/FUTURE EVENTS	6. ART/MUSIC/SPORTS

HOW ARE THE PROBLEMS SOLVED

EXAMPLE 1	EXAMPLE 2
EXAMPLE 3	EXAMPLE 4

TWO WORDS FROM THE LESSON TO CONTRAST/COMPARE

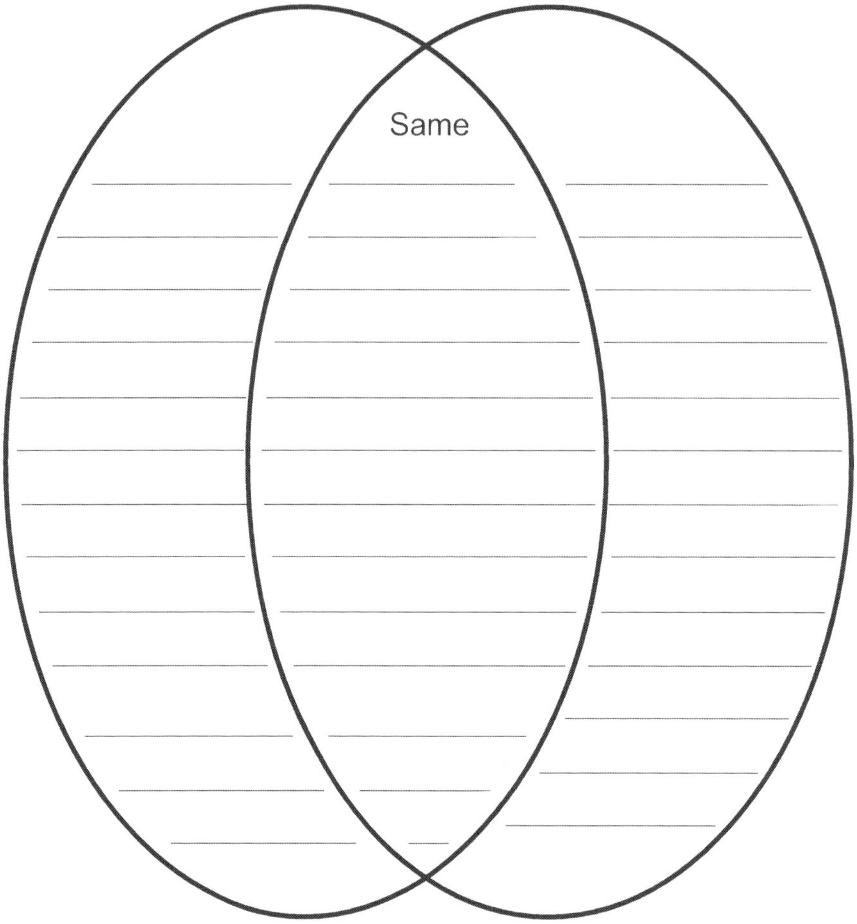

STEPS TAUGHT IN THIS

LESSON

1 ____________________

2 ____________________

3 ____________________

MATH WORK OR SYNONYM/ ANTONYM

1 ____________________

2 ____________________

3 ____________________

WORD	DETAIL
1 ______	1 ______
2 ______	2 ______
3 ______	3 ______
4 ______	4 ______
5 ______	5 ______
6 ______	6 ______
7 ______	7 ______
8 ______	8 ______
9 ______	9 ______
10 ______	10 ______

DEFINITION OR KEY WORDS

PICTURE OR EXAMPLE

CONCEPT OR SKILL

STRATEGIES

COMMON MISTAKES OR MISCONCEPTIONS

HOW DOES THIS RELATE TO OTHER MATH TOPICS?

______________________ ______________________

______________________ ______________________

Math Action Terms/Words

Formulas	Equations

EXIT TICKET: THREE THINGS I LEARNED/WHAT STOOD OUT

1	
2	
3	

NEED PRACTICE	ALMOST GOT IT	GOOD AT THIS SKILL

TEACHER NOTE.

PARENT NOTE

DATE ____________

CONCEPT OR SKILL

____

WHAT JOB USES THIS MATH?

____

ASK TEACHER WHERE IN THE NEIGHBORHOOD, HOME, MONEY, NATURE, PAST/FUTURE EVENTS, AND/OR, ARTS/MUSIC/SPORTS WILL YOU SEE OR USE THIS SKILL?

REAL LIFE CONNECTION- ASK TEACHER WHERE WILL SEE THIS SKILL IN...

1. NEIGHBORHOOD	2. HOME	3. MONEY
4. NATURE	5. PAST/FUTURE EVENTS	6. ART/MUSIC/SPORTS

HOW ARE THE PROBLEMS SOLVED

EXAMPLE 1	EXAMPLE 2
EXAMPLE 3	EXAMPLE 4

TWO WORDS FROM THE LESSON TO CONTRAST/COMPARE

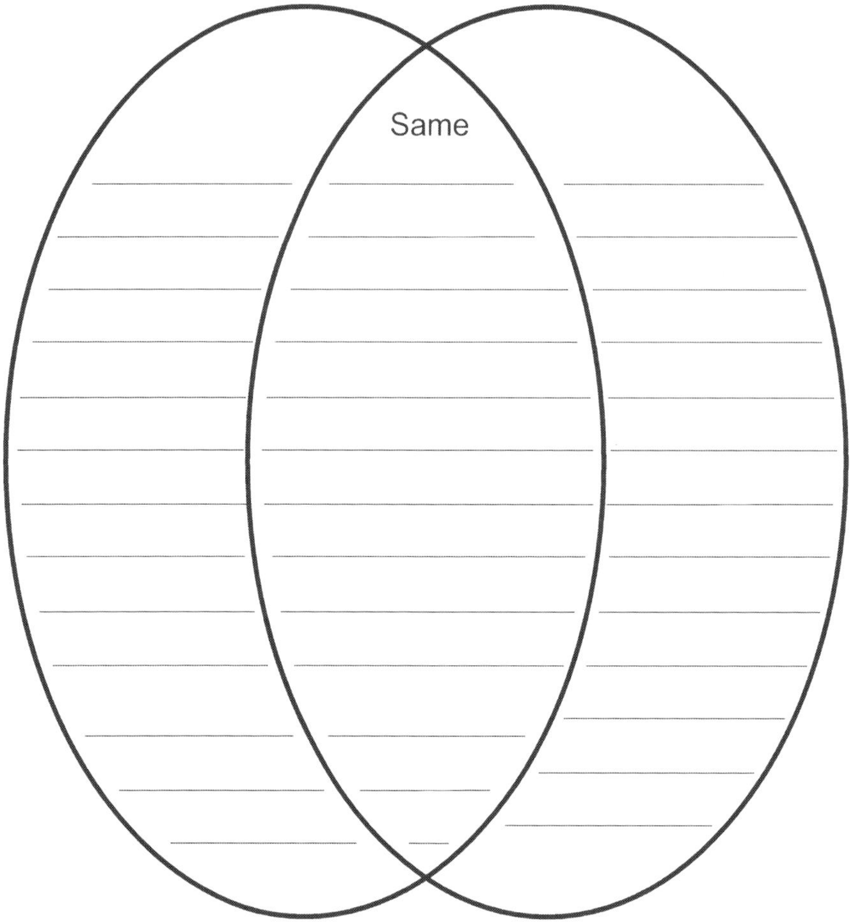

STEPS TAUGHT IN THIS LESSON

1 ______

2 ______

3 ______

MATH WORK OR SYNONYM/ ANTONYM

1 ______

2 ______

3 ______

WORD	DETAIL
1 ______	1 ______
2 ______	2 ______
3 ______	3 ______
4 ______	4 ______
5 ______	5 ______
6 ______	6 ______
7 ______	7 ______
8 ______	8 ______
9 ______	9 ______
10 ______	10 ______

DEFINITION OR KEY WORDS

PICTURE OR EXAMPLE

CONCEPT OR SKILL

STRATEGIES

COMMON MISTAKES OR MISCONCEPTIONS

HOW DOES THIS RELATE TO OTHER MATH TOPICS?

_______________ _______________

_______________ _______________

Math Action Terms/Words

Formulas	Equations

EXIT TICKET: THREE THINGS I LEARNED/WHAT STOOD OUT

1
2
3

NEED PRACTICE	ALMOST GOT IT	GOOD AT THIS SKILL

TEACHER NOTE.

PARENT NOTE

DATE ____________

CONCEPT OR SKILL

____

WHAT JOB USES THIS MATH?

____

ASK TEACHER WHERE IN THE NEIGHBORHOOD, HOME, MONEY, NATURE, PAST/FUTURE EVENTS, AND/OR, ARTS/MUSIC/SPORTS WILL YOU SEE OR USE THIS SKILL?

REAL LIFE CONNECTION-ASK TEACHER WHERE WILL SEE THIS SKILL IN:

1. NEIGHBORHOOD	2. HOME	3. MONEY
4. NATURE	5. PAST/FUTURE EVENTS	6. ART/MUSIC/SPORTS

HOW ARE THE PROBLEMS SOLVED

EXAMPLE 1	EXAMPLE 2
EXAMPLE 3	EXAMPLE 4

TWO WORDS FROM THE LESSON TO CONTRAST/COMPARE

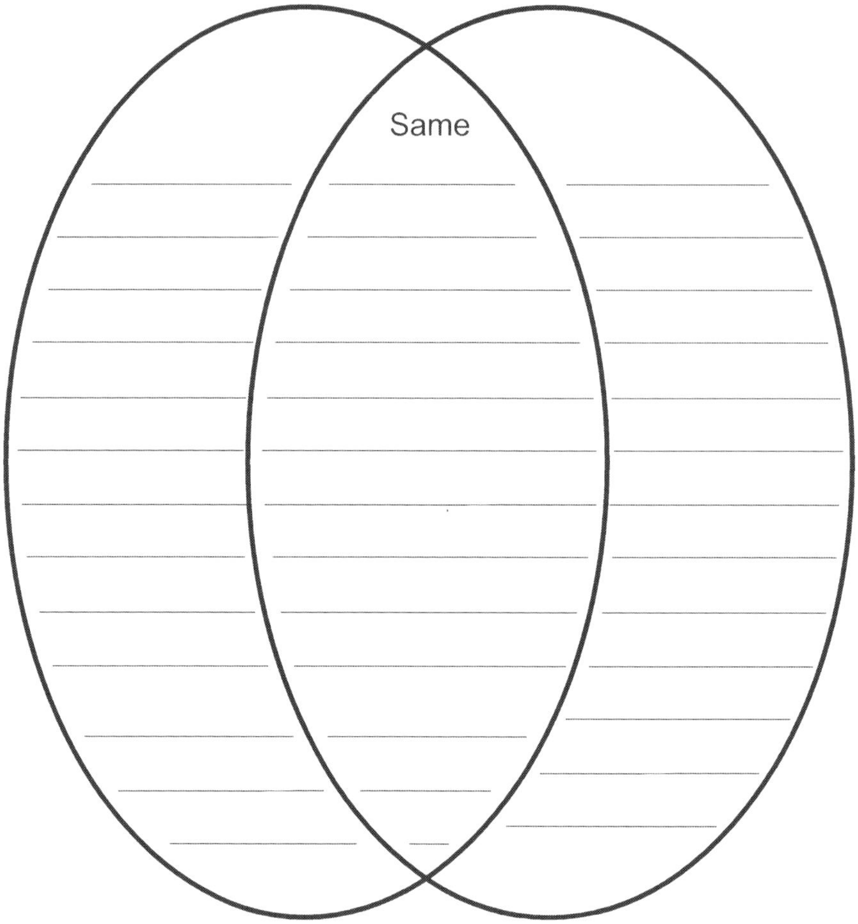

STEPS TAUGHT IN THIS LESSON

1 ______

2 ______

3 ______

MATH WORK OR SYNONYM/ ANTONYM

1 ______

2 ______

3 ______

WORD	DETAIL
1 ______	1 ______
2 ______	2 ______
3 ______	3 ______
4 ______	4 ______
5 ______	5 ______
6 ______	6 ______
7 ______	7 ______
8 ______	8 ______
9 ______	9 ______
10 ______	10 ______

DEFINITION OR KEY WORDS

PICTURE OR EXAMPLE

CONCEPT OR SKILL

STRATEGIES

COMMON MISTAKES OR MISCONCEPTIONS

HOW DOES THIS RELATE TO OTHER MATH TOPICS?

______________________ ______________________

______________________ ______________________

Math Action Terms/Words

Formulas	Equations

EXIT TICKET: THREE THINGS I LEARNED/WHAT STOOD OUT

1	
2	
3	

NEED PRACTICE	ALMOST GOT IT	GOOD AT THIS SKILL

TEACHER NOTE.

PARENT NOTE

DATE ______________

CONCEPT OR SKILL

____

WHAT JOB USES THIS MATH?

____

ASK TEACHER WHERE IN THE NEIGHBORHOOD, HOME, MONEY, NATURE, PAST/FUTURE EVENTS, AND/OR, ARTS/MUSIC/SPORTS WILL YOU SEE OR USE THIS SKILL?

1. NEIGHBORHOOD	2. HOME	3. MONEY
4. NATURE	5. PAST/FUTURE EVENTS	6. ART/MUSIC/SPORTS

HOW ARE THE PROBLEMS SOLVED

EXAMPLE 1	EXAMPLE 2
EXAMPLE 3	EXAMPLE 4

TWO WORDS FROM THE LESSON TO CONTRAST/COMPARE

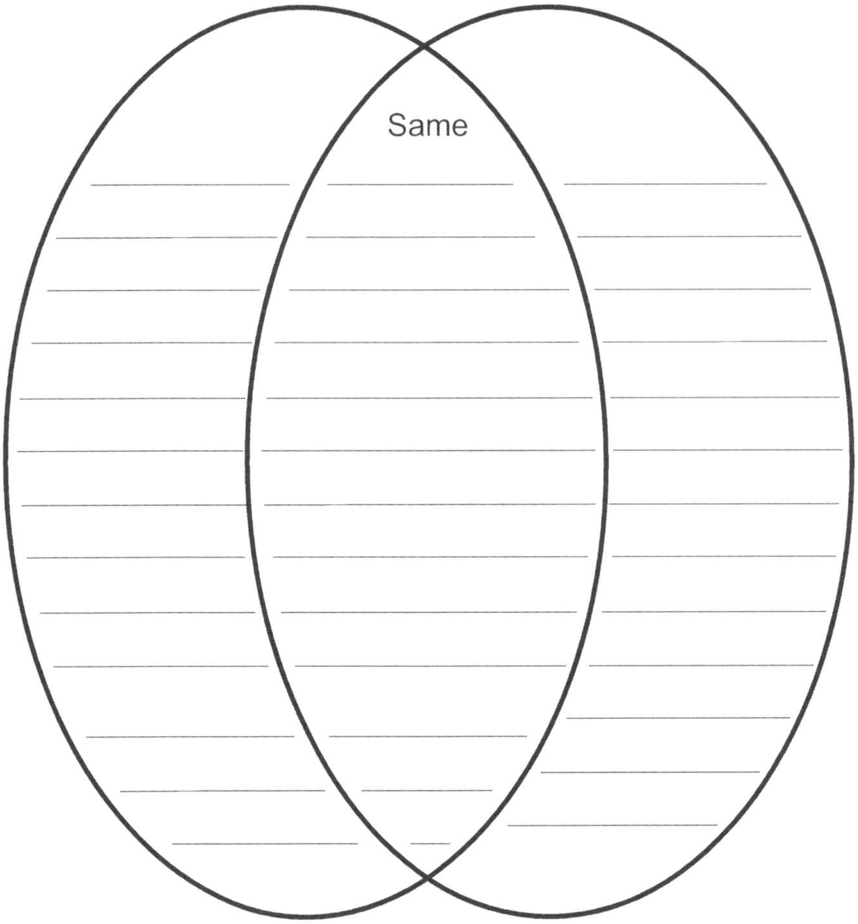

STEPS TAUGHT IN THIS LESSON

1 ______

2 ______

3 ______

MATH WORK OR SYNONYM/ ANTONYM

1 ______

2 ______

3 ______

WORD	DETAIL
1 ______	1 ______
2 ______	2 ______
3 ______	3 ______
4 ______	4 ______
5 ______	5 ______
6 ______	6 ______
7 ______	7 ______
8 ______	8 ______
9 ______	9 ______
10 ______	10 ______

DEFINITION OR KEY WORDS

PICTURE OR EXAMPLE

CONCEPT OR SKILL

STRATEGIES

COMMON MISTAKES OR MISCONCEPTIONS

HOW DOES THIS RELATE TO OTHER MATH TOPICS?

______________________ ______________________

______________________ ______________________

Math Action Terms/Words

Formulas	Equations

EXIT TICKET: THREE THINGS I LEARNED/WHAT STOOD OUT

1	
2	
3	

NEED PRACTICE	ALMOST GOT IT	GOOD AT THIS SKILL

TEACHER NOTE.

PARENT NOTE

DATE ____________

CONCEPT OR SKILL

____

WHAT JOB USES THIS MATH?

____

ASK TEACHER WHERE IN THE NEIGHBORHOOD, HOME, MONEY, NATURE, PAST/FUTURE EVENTS, AND/OR, ARTS/MUSIC/SPORTS WILL YOU SEE OR USE THIS SKILL?

1. NEIGHBORHOOD	2. HOME	3. MONEY
4. NATURE	5. PAST/FUTURE EVENTS	6. ART/MUSIC/SPORTS

HOW ARE THE PROBLEMS SOLVED

EXAMPLE 1	EXAMPLE 2
EXAMPLE 3	EXAMPLE 4

TWO WORDS FROM THE LESSON TO CONTRAST/COMPARE

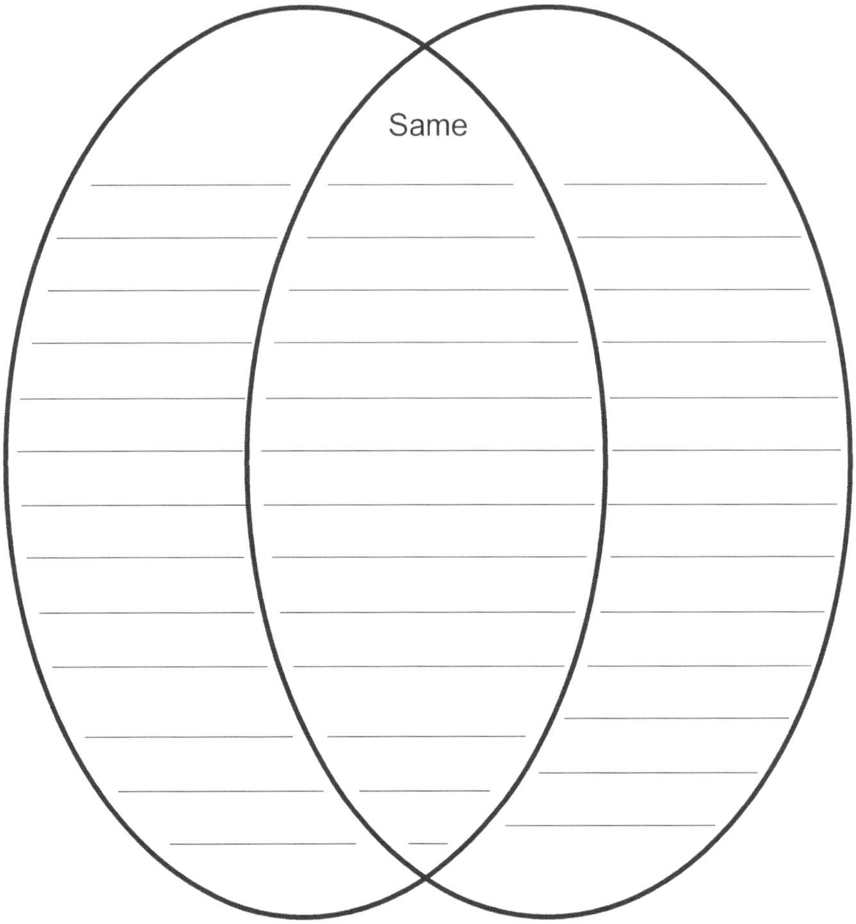

STEPS TAUGHT IN THIS LESSON

1 ____________________

2 ____________________

3 ____________________

MATH WORK OR SYNONYM/ ANTONYM

1 ____________________

2 ____________________

3 ____________________

WORD	DETAIL
1	1
2	2
3	3
4	4
5	5
6	6
7	7
8	8
9	9
10	10

DEFINITION OR KEY WORDS

PICTURE OR EXAMPLE

CONCEPT OR SKILL

STRATEGIES

COMMON MISTAKES OR MISCONCEPTIONS

HOW DOES THIS RELATE TO OTHER MATH TOPICS?

Math Action Terms/Words

Formulas	Equations

EXIT TICKET: THREE THINGS I LEARNED/WHAT STOOD OUT

1
2
3

NEED PRACTICE	ALMOST GOT IT	GOOD AT THIS SKILL

TEACHER NOTE.	PARENT NOTE

DATE ____________

CONCEPT OR SKILL

____

WHAT JOB USES THIS MATH?

____

ASK TEACHER WHERE IN THE NEIGHBORHOOD, HOME, MONEY, NATURE, PAST/FUTURE EVENTS, AND/OR, ARTS/MUSIC/SPORTS WILL YOU SEE OR USE THIS SKILL?

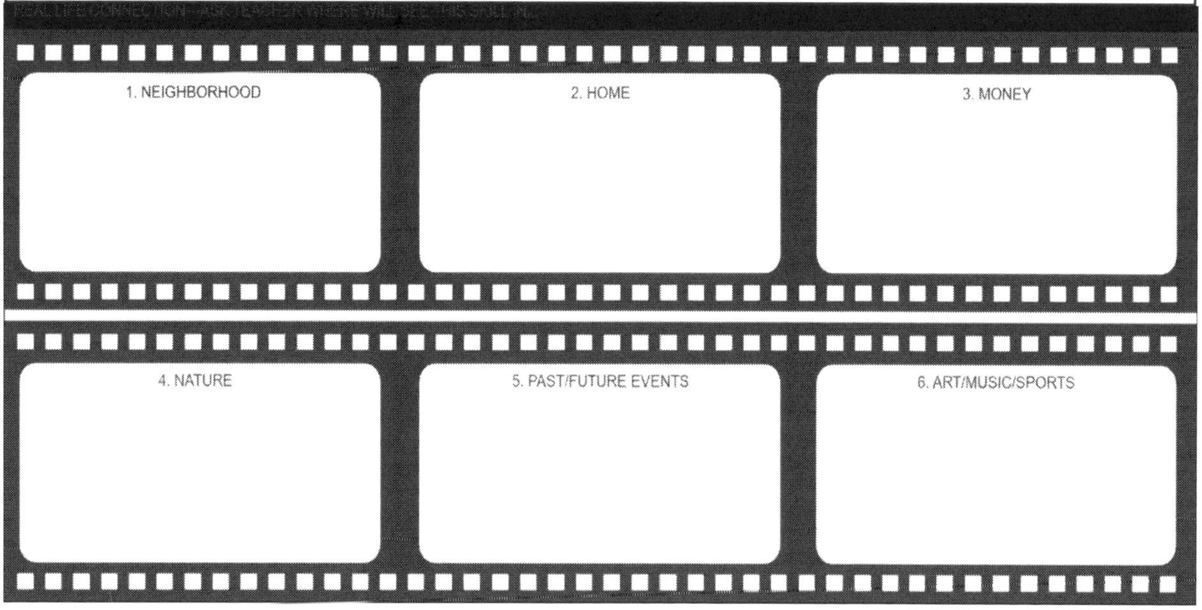

HOW ARE THE PROBLEMS SOLVED

EXAMPLE 1	EXAMPLE 2
EXAMPLE 3	EXAMPLE 4

TWO WORDS FROM THE LESSON TO CONTRAST/COMPARE

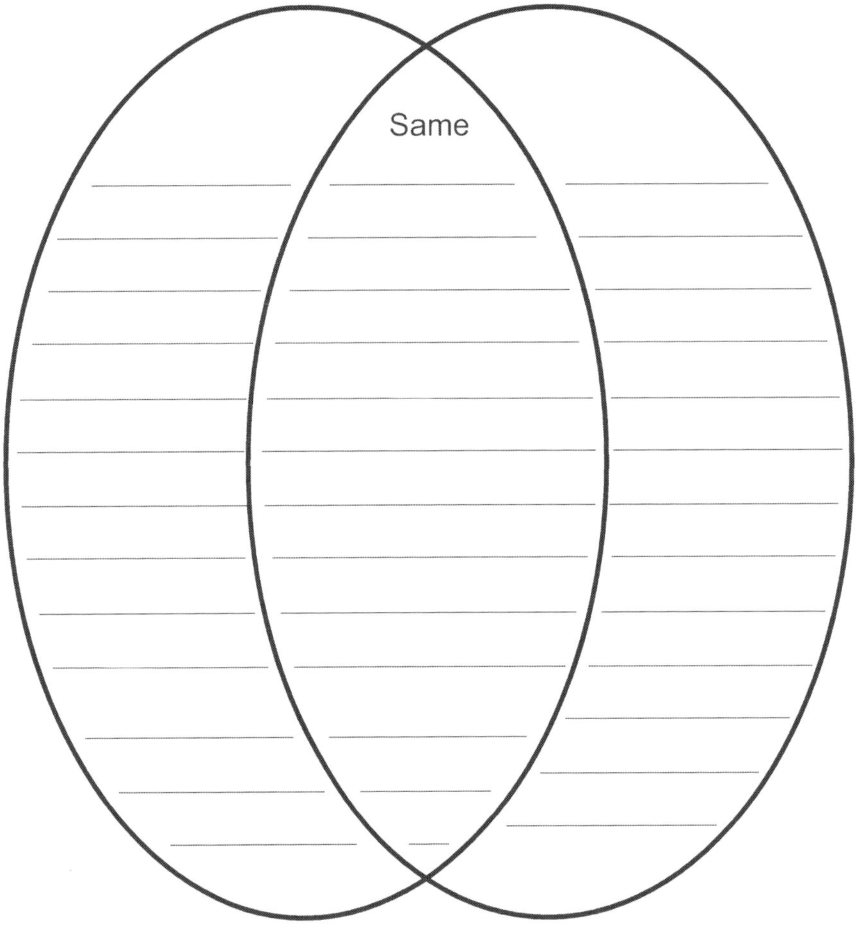

STEPS TAUGHT IN THIS

LESSON

1 ____________________

2 ____________________

3 ____________________

MATH WORK OR SYNONYM/ ANTONYM

1 ____________________

2 ____________________

3 ____________________

WORD	DETAIL
1 ____________	1 ____________
2 ____________	2 ____________
3 ____________	3 ____________
4 ____________	4 ____________
5 ____________	5 ____________
6 ____________	6 ____________
7 ____________	7 ____________
8 ____________	8 ____________
9 ____________	9 ____________
10 ____________	10 ____________

DEFINITION OR KEY WORDS

PICTURE OR EXAMPLE

CONCEPT OR SKILL

STRATEGIES

COMMON MISTAKES OR MISCONCEPTIONS

HOW DOES THIS RELATE TO OTHER MATH TOPICS?

Math Action Terms/Words

Formulas	Equations

EXIT TICKET: THREE THINGS I LEARNED/WHAT STOOD OUT

1	
2	
3	

NEED PRACTICE	ALMOST GOT IT	GOOD AT THIS SKILL

TEACHER NOTE.

PARENT NOTE

DATE ____________

CONCEPT OR SKILL

____

WHAT JOB USES THIS MATH?

____

ASK TEACHER WHERE IN THE NEIGHBORHOOD, HOME, MONEY, NATURE, PAST/FUTURE EVENTS, AND/OR, ARTS/MUSIC/SPORTS WILL YOU SEE OR USE THIS SKILL?

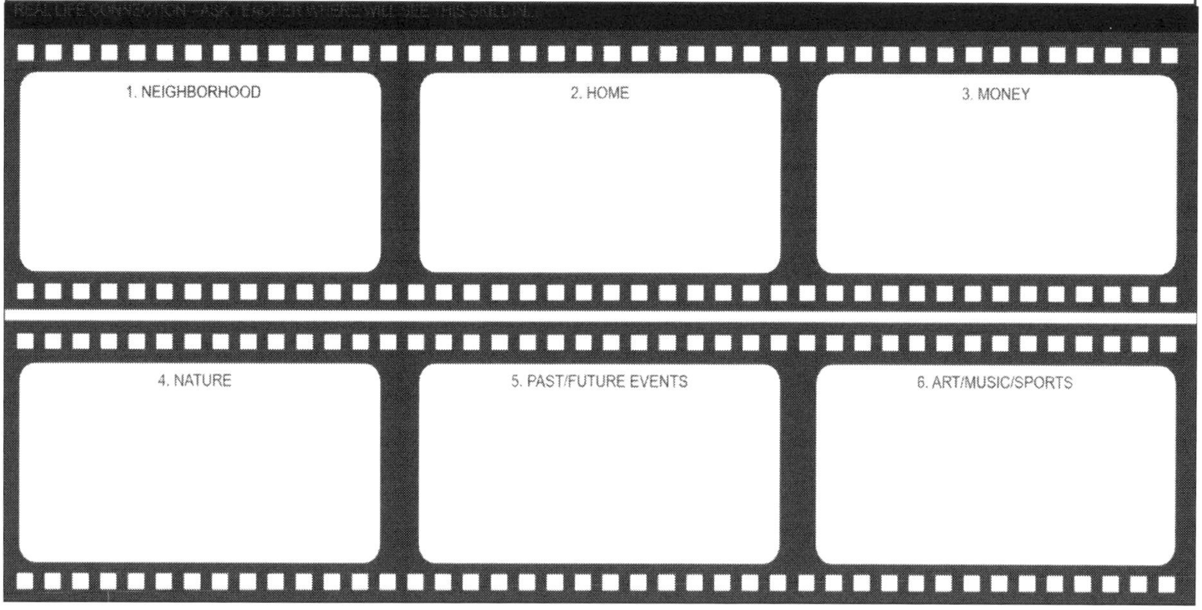

HOW ARE THE PROBLEMS SOLVED

EXAMPLE 1	EXAMPLE 2
EXAMPLE 3	EXAMPLE 4

TWO WORDS FROM THE LESSON TO CONTRAST/COMPARE

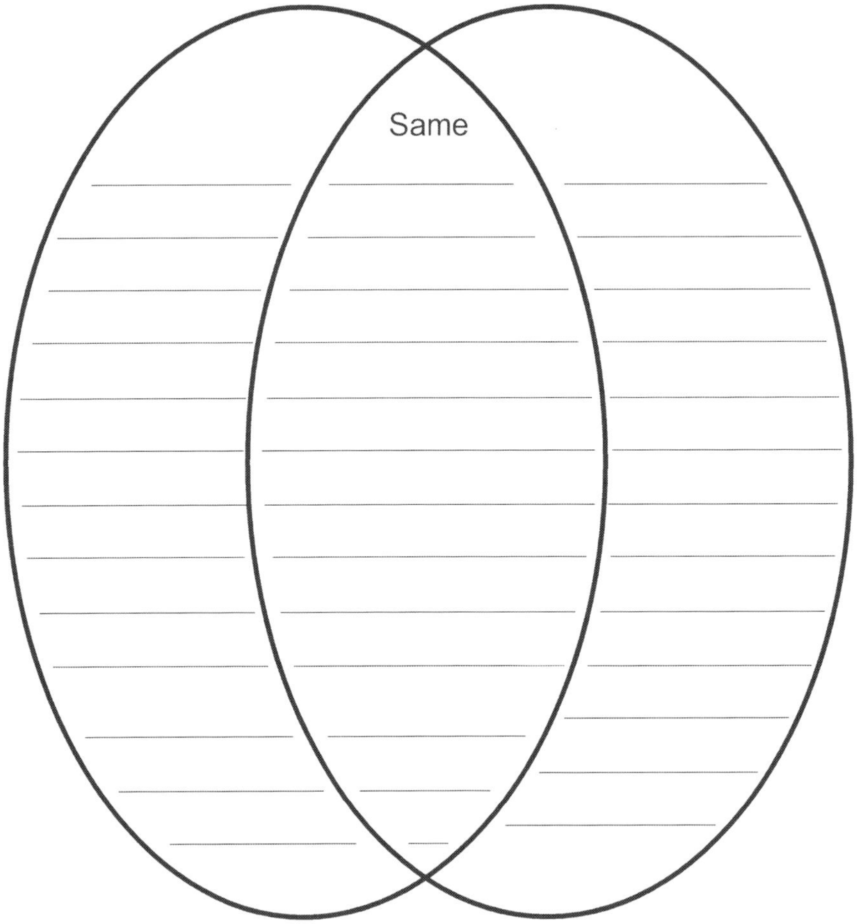

STEPS TAUGHT IN THIS

LESSON

1 __________

2 __________

3 __________

MATH WORK OR SYNONYM/ ANTONYM

1 __________

2 __________

3 __________

WORD	DETAIL
1	1
2	2
3	3
4	4
5	5
6	6
7	7
8	8
9	9
10	10

DEFINITION OR KEY WORDS

PICTURE OR EXAMPLE

CONCEPT OR SKILL

STRATEGIES

COMMON MISTAKES OR MISCONCEPTIONS

HOW DOES THIS RELATE TO OTHER MATH TOPICS?

___________________________ ___________________________

___________________________ ___________________________

Math Action Terms/Words

Formulas	Equations

EXIT TICKET: THREE THINGS I LEARNED/WHAT STOOD OUT

1	
2	
3	

NEED PRACTICE	ALMOST GOT IT	GOOD AT THIS SKILL

TEACHER NOTE.

PARENT NOTE

DATE ____________

CONCEPT OR SKILL

____

WHAT JOB USES THIS MATH?

____

ASK TEACHER WHERE IN THE NEIGHBORHOOD, HOME, MONEY, NATURE, PAST/FUTURE EVENTS, AND/OR, ARTS/MUSIC/SPORTS WILL YOU SEE OR USE THIS SKILL?

REAL LIFE CONNECTION – ASK TEACHER WHERE WILL SEE THIS SKILL IN:

1. NEIGHBORHOOD	2. HOME	3. MONEY
4. NATURE	5. PAST/FUTURE EVENTS	6. ART/MUSIC/SPORTS

HOW ARE THE PROBLEMS SOLVED

EXAMPLE 1	EXAMPLE 2
EXAMPLE 3	EXAMPLE 4

TWO WORDS FROM THE LESSON TO CONTRAST/COMPARE

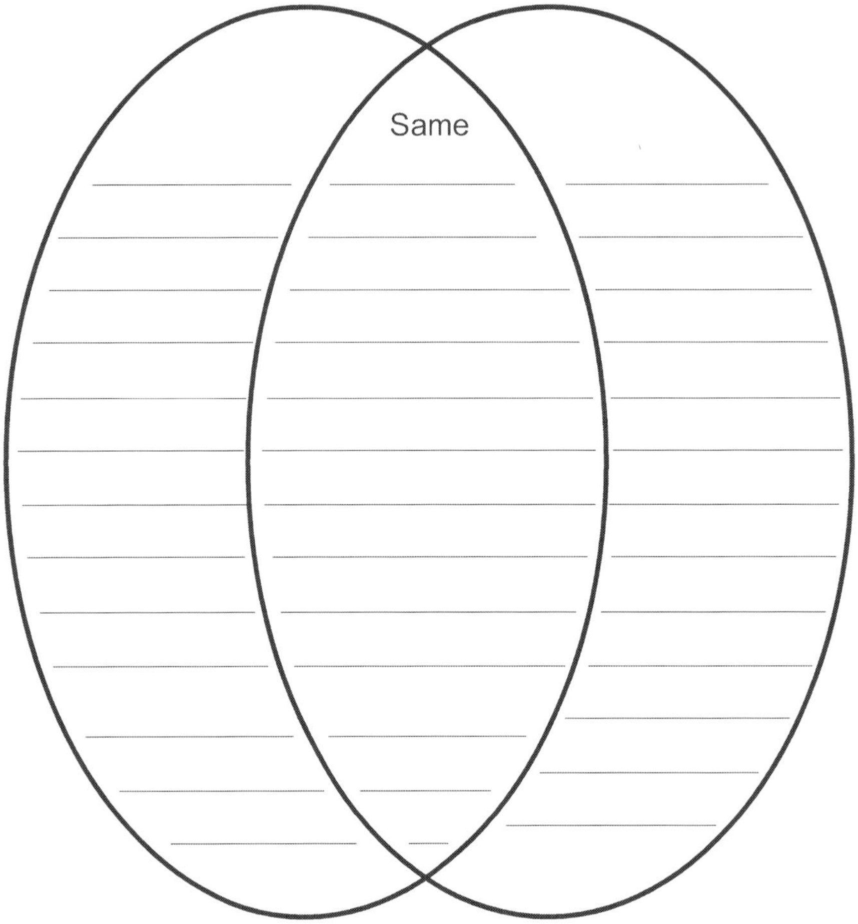

STEPS TAUGHT IN THIS

LESSON

1 ______

2 ______

3 ______

MATH WORK OR SYNONYM/ ANTONYM

1 ______

2 ______

3 ______

WORD	DETAIL
1 ______	1 ______
2 ______	2 ______
3 ______	3 ______
4 ______	4 ______
5 ______	5 ______
6 ______	6 ______
7 ______	7 ______
8 ______	8 ______
9 ______	9 ______
10 ______	10 ______

DEFINITION OR KEY WORDS

PICTURE OR EXAMPLE

CONCEPT OR SKILL

STRATEGIES

COMMON MISTAKES OR MISCONCEPTIONS

HOW DOES THIS RELATE TO OTHER MATH TOPICS?

____________________ ____________________

____________________ ____________________

Math Action Terms/Words

Formulas	Equations

EXIT TICKET: THREE THINGS I LEARNED/WHAT STOOD OUT

1	
2	
3	

NEED PRACTICE	ALMOST GOT IT	GOOD AT THIS SKILL

TEACHER NOTE. PARENT NOTE

DATE ____________

CONCEPT OR SKILL

____

WHAT JOB USES THIS MATH?

____

ASK TEACHER WHERE IN THE NEIGHBORHOOD, HOME, MONEY, NATURE, PAST/FUTURE EVENTS, AND/OR, ARTS/MUSIC/SPORTS WILL YOU SEE OR USE THIS SKILL?

REAL LIFE CONNECTION - ASK TEACHER WHERE WILL SEE THIS SKILL IN.

1. NEIGHBORHOOD	2. HOME	3. MONEY
4. NATURE	5. PAST/FUTURE EVENTS	6. ART/MUSIC/SPORTS

HOW ARE THE PROBLEMS SOLVED

EXAMPLE 1	EXAMPLE 2
EXAMPLE 3	EXAMPLE 4

TWO WORDS FROM THE LESSON TO CONTRAST/COMPARE

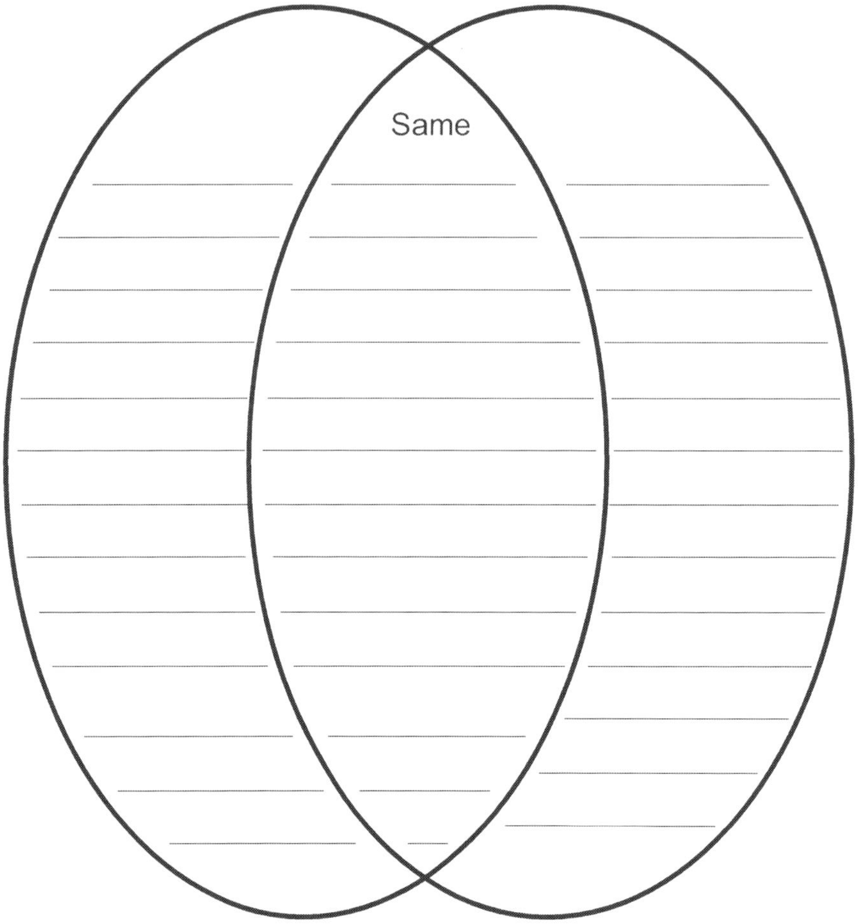

STEPS TAUGHT IN THIS LESSON

1

2

3

MATH WORK OR SYNONYM/ ANTONYM

1

2

3

WORD	DETAIL
1	1
2	2
3	3
4	4
5	5
6	6
7	7
8	8
9	9
10	10

DEFINITION OR KEY WORDS

PICTURE OR EXAMPLE

CONCEPT OR SKILL

STRATEGIES

COMMON MISTAKES OR MISCONCEPTIONS

HOW DOES THIS RELATE TO OTHER MATH TOPICS?

_______________ _______________

_______________ _______________

Math Action Terms/Words

Formulas	Equations

EXIT TICKET: THREE THINGS I LEARNED/WHAT STOOD OUT

1	
2	
3	

NEED PRACTICE	ALMOST GOT IT	GOOD AT THIS SKILL

TEACHER NOTE. PARENT NOTE

DATE ____________

CONCEPT OR SKILL

____

WHAT JOB USES THIS MATH?

____

ASK TEACHER WHERE IN THE NEIGHBORHOOD, HOME, MONEY, NATURE, PAST/FUTURE EVENTS, AND/OR, ARTS/MUSIC/SPORTS WILL YOU SEE OR USE THIS SKILL?

1. NEIGHBORHOOD	2. HOME	3. MONEY
4. NATURE	5. PAST/FUTURE EVENTS	6. ART/MUSIC/SPORTS

HOW ARE THE PROBLEMS SOLVED

EXAMPLE 1	EXAMPLE 2
EXAMPLE 3	EXAMPLE 4

TWO WORDS FROM THE LESSON TO CONTRAST/COMPARE

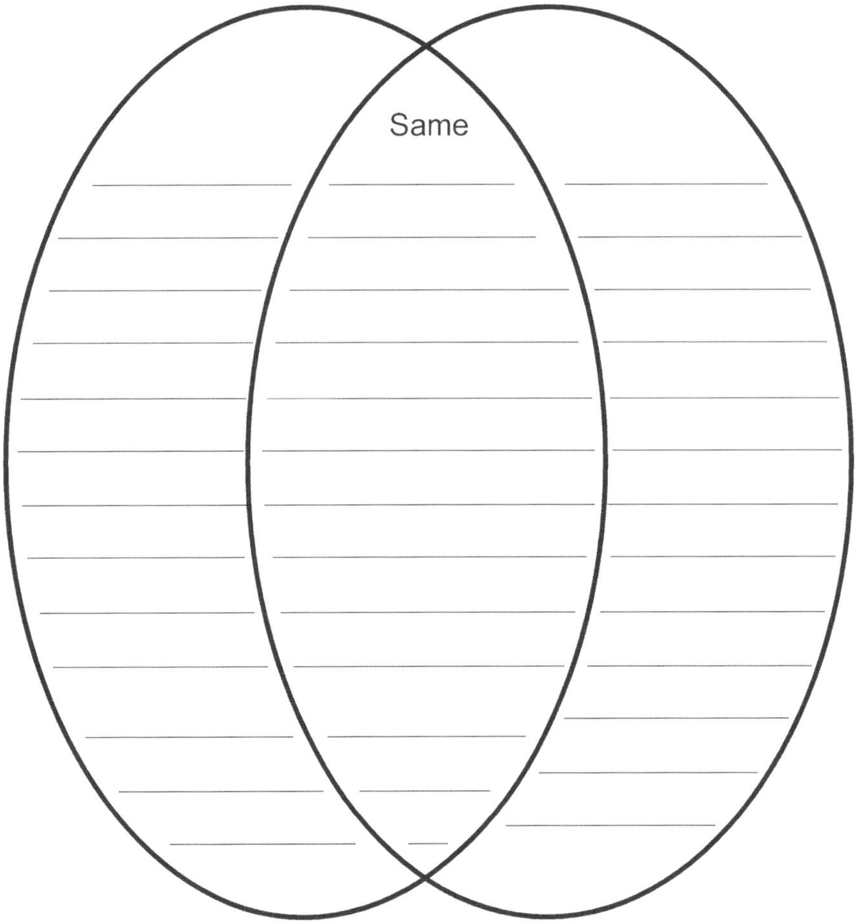

STEPS TAUGHT IN THIS

LESSON

1 ______________________

2 ______________________

3 ______________________

MATH WORK OR SYNONYM/ ANTONYM

1 ______________________

2 ______________________

3 ______________________

WORD	DETAIL
1 ______________	1 ______________
2 ______________	2 ______________
3 ______________	3 ______________
4 ______________	4 ______________
5 ______________	5 ______________
6 ______________	6 ______________
7 ______________	7 ______________
8 ______________	8 ______________
9 ______________	9 ______________
10 ______________	10 ______________

DEFINITION OR KEY WORDS

PICTURE OR EXAMPLE

CONCEPT OR SKILL

STRATEGIES

COMMON MISTAKES OR MISCONCEPTIONS

HOW DOES THIS RELATE TO OTHER MATH TOPICS?

______________________ ______________________

______________________ ______________________

Math Action Terms/Words

Formulas	Equations

EXIT TICKET: THREE THINGS I LEARNED/WHAT STOOD OUT

1	
2	
3	

NEED PRACTICE	ALMOST GOT IT	GOOD AT THIS SKILL

TEACHER NOTE.

PARENT NOTE

DATE ____________

CONCEPT OR SKILL

____

WHAT JOB USES THIS MATH?

____

ASK TEACHER WHERE IN THE NEIGHBORHOOD, HOME, MONEY, NATURE, PAST/FUTURE EVENTS, AND/OR, ARTS/MUSIC/SPORTS WILL YOU SEE OR USE THIS SKILL?

REAL LIFE CONNECTION—ASK TEACHER WHERE WILL SEE THIS SKILL IN:

1. NEIGHBORHOOD	2. HOME	3. MONEY
4. NATURE	5. PAST/FUTURE EVENTS	6. ART/MUSIC/SPORTS

HOW ARE THE PROBLEMS SOLVED

EXAMPLE 1	EXAMPLE 2
EXAMPLE 3	EXAMPLE 4

TWO WORDS FROM THE LESSON TO CONTRAST/COMPARE

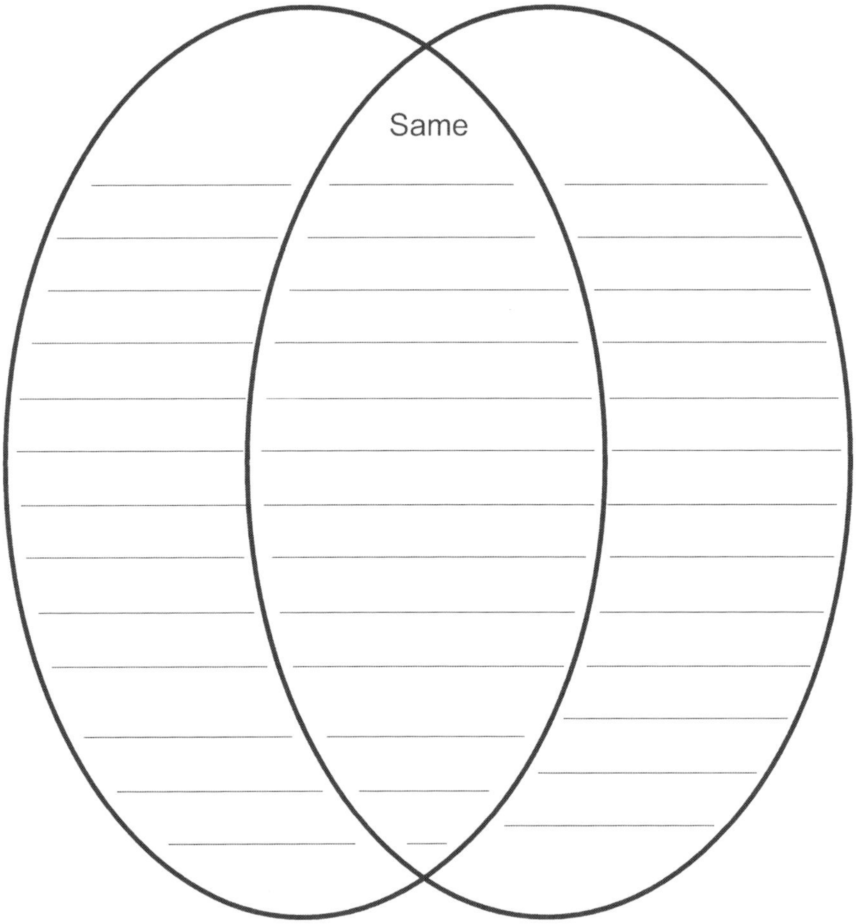

STEPS TAUGHT IN THIS LESSON

1 ______

2 ______

3 ______

MATH WORK OR SYNONYM/ ANTONYM

1 ______

2 ______

3 ______

WORD	DETAIL
1 ______	1 ______
2 ______	2 ______
3 ______	3 ______
4 ______	4 ______
5 ______	5 ______
6 ______	6 ______
7 ______	7 ______
8 ______	8 ______
9 ______	9 ______
10 ______	10 ______

DEFINITION OR KEY WORDS

PICTURE OR EXAMPLE

CONCEPT OR SKILL

STRATEGIES

COMMON MISTAKES OR MISCONCEPTIONS

HOW DOES THIS RELATE TO OTHER MATH TOPICS?

______________________ ______________________

______________________ ______________________

Math Action Terms/Words

Formulas	Equations

EXIT TICKET: THREE THINGS I LEARNED/WHAT STOOD OUT

1	
2	
3	

NEED PRACTICE	ALMOST GOT IT	GOOD AT THIS SKILL

TEACHER NOTE.	PARENT NOTE

DATE ____________

CONCEPT OR SKILL

____

WHAT JOB USES THIS MATH?

____

ASK TEACHER WHERE IN THE NEIGHBORHOOD, HOME, MONEY, NATURE, PAST/FUTURE EVENTS, AND/OR, ARTS/MUSIC/SPORTS WILL YOU SEE OR USE THIS SKILL?

REAL LIFE CONNECTION – ASK TEACHER WHERE I WILL SEE THIS SKILL IN...

1. NEIGHBORHOOD	2. HOME	3. MONEY

4. NATURE	5. PAST/FUTURE EVENTS	6. ART/MUSIC/SPORTS

HOW ARE THE PROBLEMS SOLVED

EXAMPLE 1	EXAMPLE 2
EXAMPLE 3	EXAMPLE 4

TWO WORDS FROM THE LESSON TO CONTRAST/COMPARE

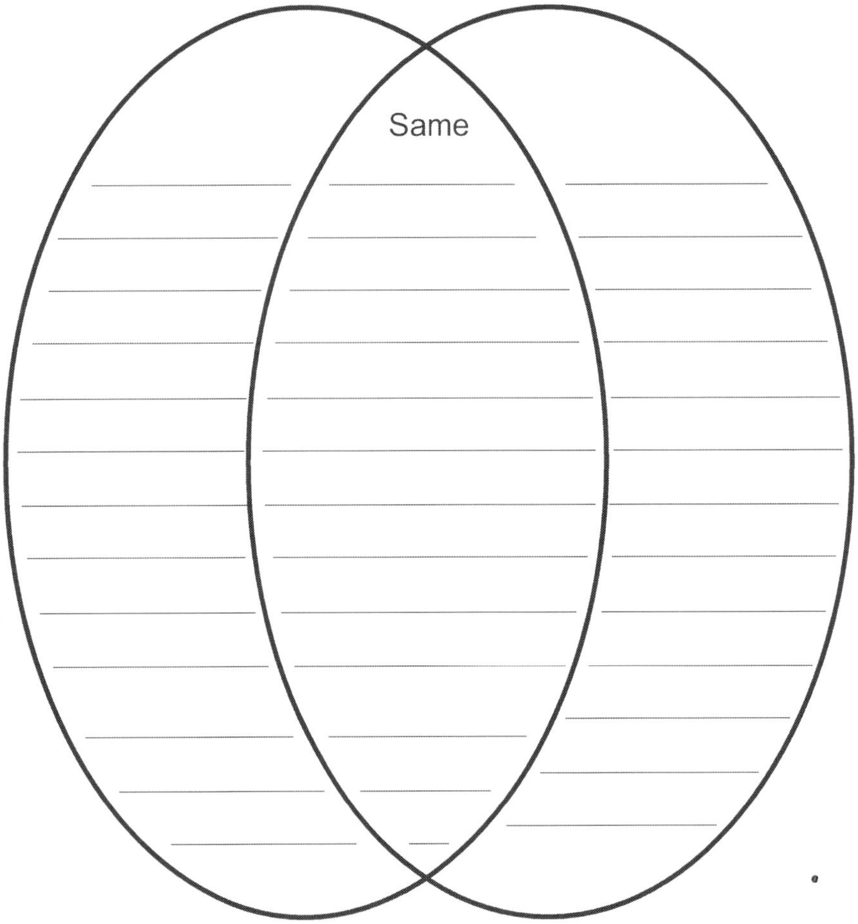

STEPS TAUGHT IN THIS

LESSON

1 ____________________

2 ____________________

3 ____________________

MATH WORK OR SYNONYM/ ANTONYM

1 ____________________

2 ____________________

3 ____________________

WORD	DETAIL
1 ______	1 ______
2 ______	2 ______
3 ______	3 ______
4 ______	4 ______
5 ______	5 ______
6 ______	6 ______
7 ______	7 ______
8 ______	8 ______
9 ______	9 ______
10 ______	10 ______

DEFINITION OR KEY WORDS

PICTURE OR EXAMPLE

CONCEPT OR SKILL

STRATEGIES

COMMON MISTAKES OR MISCONCEPTIONS

HOW DOES THIS RELATE TO OTHER MATH TOPICS?

______________________ ______________________

______________________ ______________________

Math Action Terms/Words

Formulas	Equations

EXIT TICKET: THREE THINGS I LEARNED/WHAT STOOD OUT

1	
2	
3	

NEED PRACTICE	ALMOST GOT IT	GOOD AT THIS SKILL

TEACHER NOTE.	PARENT NOTE

DATE ____________

CONCEPT OR SKILL

____

WHAT JOB USES THIS MATH?

____

ASK TEACHER WHERE IN THE NEIGHBORHOOD, HOME, MONEY, NATURE, PAST/FUTURE EVENTS, AND/OR, ARTS/MUSIC/SPORTS WILL YOU SEE OR USE THIS SKILL?

1. NEIGHBORHOOD	2. HOME	3. MONEY
4. NATURE	5. PAST/FUTURE EVENTS	6. ART/MUSIC/SPORTS

HOW ARE THE PROBLEMS SOLVED

EXAMPLE 1	EXAMPLE 2
EXAMPLE 3	EXAMPLE 4

TWO WORDS FROM THE LESSON TO CONTRAST/COMPARE

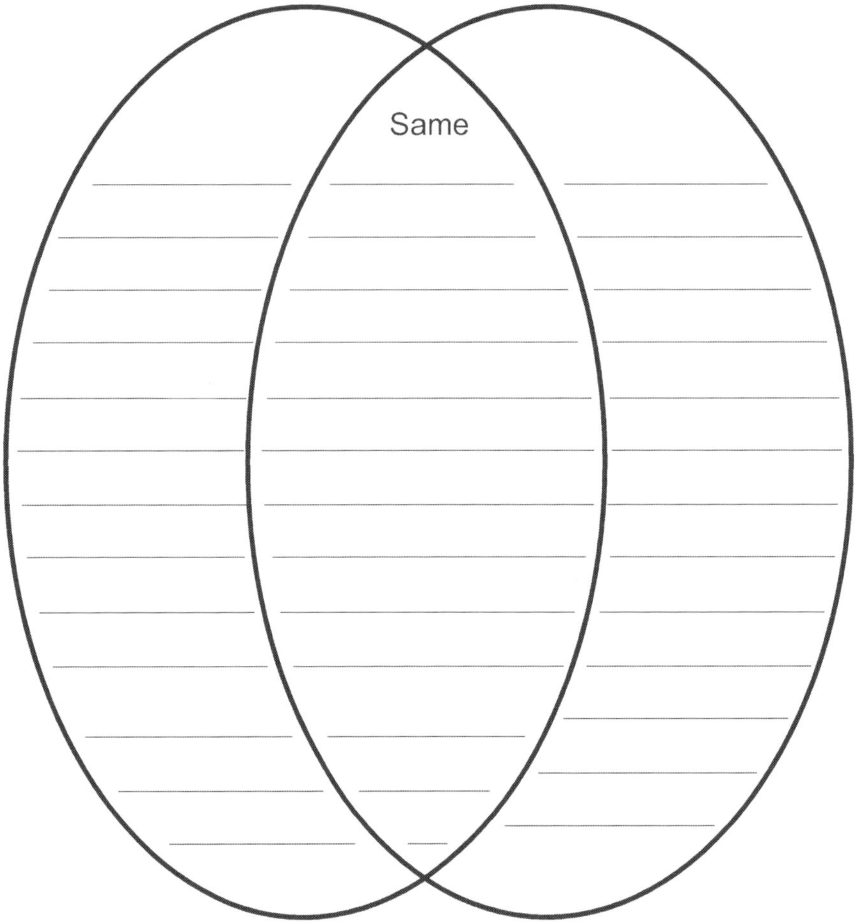

STEPS TAUGHT IN THIS LESSON

1 ____________________

2 ____________________

3 ____________________

MATH WORK OR SYNONYM/ ANTONYM

1 ____________________

2 ____________________

3 ____________________

WORD	DETAIL
1 ________	1 ________
2 ________	2 ________
3 ________	3 ________
4 ________	4 ________
5 ________	5 ________
6 ________	6 ________
7 ________	7 ________
8 ________	8 ________
9 ________	9 ________
10 ________	10 ________

DEFINITION OR KEY WORDS

PICTURE OR EXAMPLE

CONCEPT OR SKILL

STRATEGIES

COMMON MISTAKES OR MISCONCEPTIONS

HOW DOES THIS RELATE TO OTHER MATH TOPICS?

Math Action Terms/Words

Formulas	Equations

EXIT TICKET: THREE THINGS I LEARNED/WHAT STOOD OUT

1	
2	
3	

NEED PRACTICE	ALMOST GOT IT	GOOD AT THIS SKILL

TEACHER NOTE.	PARENT NOTE

DATE ____________

CONCEPT OR SKILL

____

WHAT JOB USES THIS MATH?

____

ASK TEACHER WHERE IN THE NEIGHBORHOOD, HOME, MONEY, NATURE, PAST/FUTURE EVENTS, AND/OR, ARTS/MUSIC/SPORTS WILL YOU SEE OR USE THIS SKILL?

REAL LIFE CONNECTION - ASK TEACHER WHERE WILL SEE THIS SKILL IN.

1. NEIGHBORHOOD	2. HOME	3. MONEY
4. NATURE	5. PAST/FUTURE EVENTS	6. ART/MUSIC/SPORTS

HOW ARE THE PROBLEMS SOLVED

EXAMPLE 1	EXAMPLE 2
EXAMPLE 3	EXAMPLE 4

TWO WORDS FROM THE LESSON TO CONTRAST/COMPARE

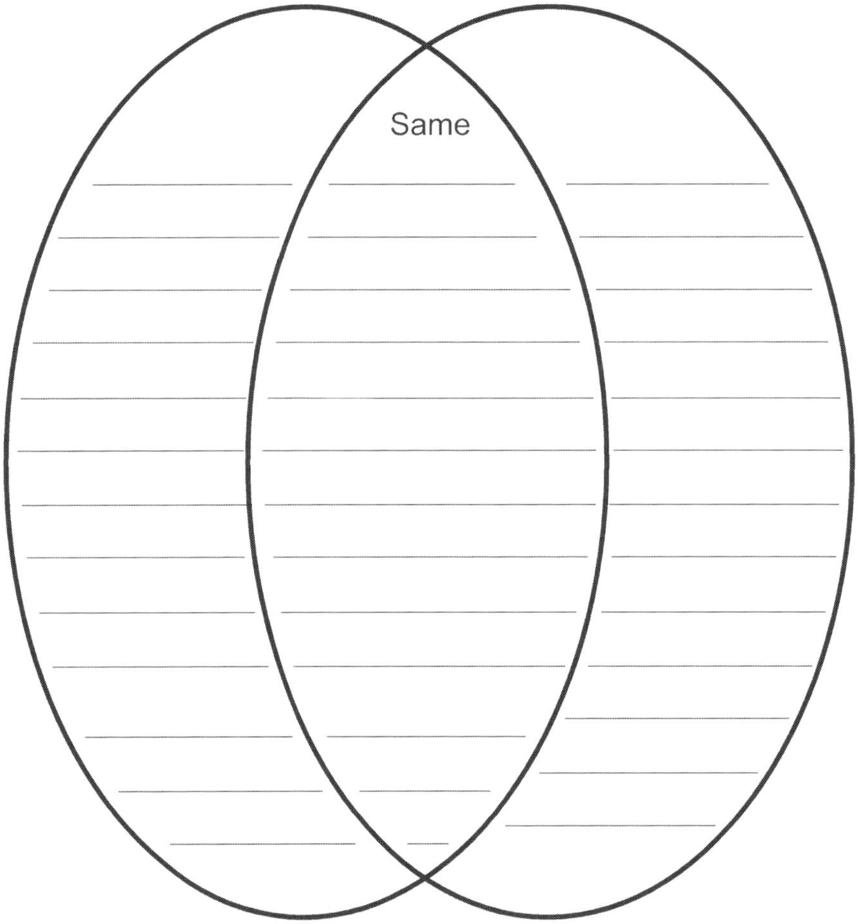

STEPS TAUGHT IN THIS

LESSON

1 ______

2 ______

3 ______

MATH WORK OR SYNONYM/ ANTONYM

1 ______

2 ______

3 ______

WORD	DETAIL
1 ______	1 ______
2 ______	2 ______
3 ______	3 ______
4 ______	4 ______
5 ______	5 ______
6 ______	6 ______
7 ______	7 ______
8 ______	8 ______
9 ______	9 ______
10 ______	10 ______

DEFINITION OR KEY WORDS

PICTURE OR EXAMPLE

CONCEPT OR SKILL

STRATEGIES

COMMON MISTAKES OR MISCONCEPTIONS

HOW DOES THIS RELATE TO OTHER MATH TOPICS?

Math Action Terms/Words

Formulas	Equations

EXIT TICKET: THREE THINGS I LEARNED/WHAT STOOD OUT

1
2
3

NEED PRACTICE	ALMOST GOT IT	GOOD AT THIS SKILL

TEACHER NOTE. PARENT NOTE

DATE ____________

CONCEPT OR SKILL

____

WHAT JOB USES THIS MATH?

____

ASK TEACHER WHERE IN THE NEIGHBORHOOD, HOME, MONEY, NATURE, PAST/FUTURE EVENTS, AND/OR, ARTS/MUSIC/SPORTS WILL YOU SEE OR USE THIS SKILL?

REAL LIFE CONNECTION - ASK TEACHER WHERE WILL SEE THIS SKILL IN...

1. NEIGHBORHOOD	2. HOME	3. MONEY
4. NATURE	5. PAST/FUTURE EVENTS	6. ART/MUSIC/SPORTS

HOW ARE THE PROBLEMS SOLVED

EXAMPLE 1	EXAMPLE 2
EXAMPLE 3	EXAMPLE 4

TWO WORDS FROM THE LESSON TO CONTRAST/COMPARE

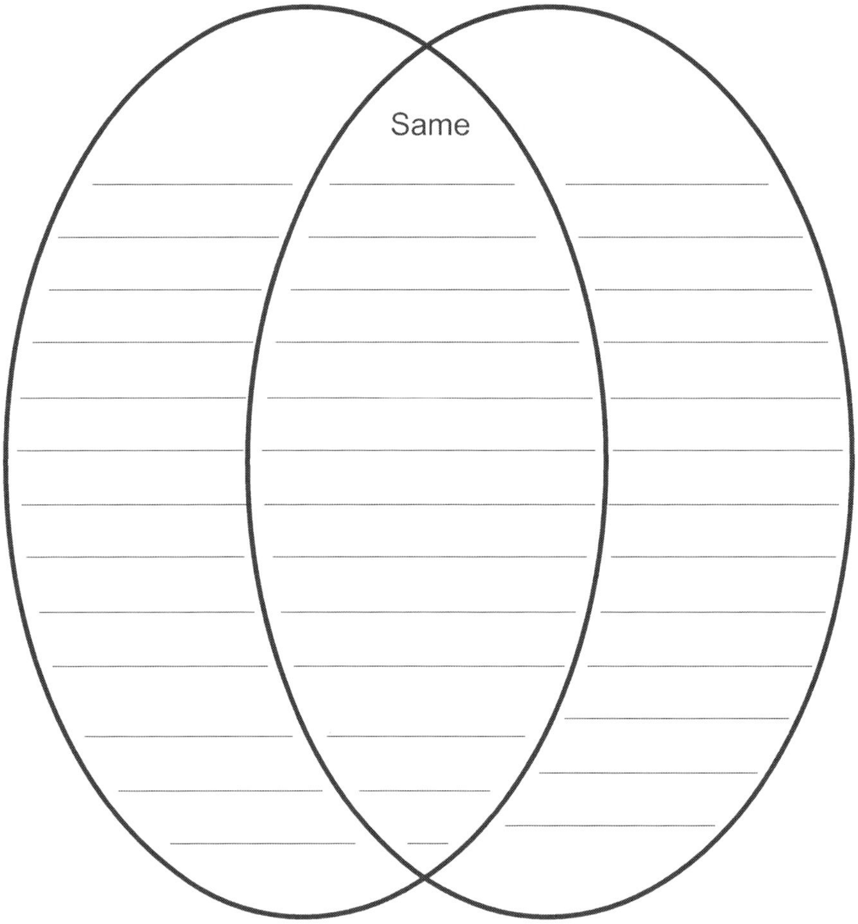

STEPS TAUGHT IN THIS LESSON

1 ____________________

2 ____________________

3 ____________________

MATH WORK OR SYNONYM/ ANTONYM

1 ____________________

2 ____________________

3 ____________________

WORD	DETAIL
1 ____________	1 ____________
2 ____________	2 ____________
3 ____________	3 ____________
4 ____________	4 ____________
5 ____________	5 ____________
6 ____________	6 ____________
7 ____________	7 ____________
8 ____________	8 ____________
9 ____________	9 ____________
10 ____________	10 ____________

DEFINITION OR KEY WORDS

PICTURE OR EXAMPLE

CONCEPT OR SKILL

STRATEGIES

COMMON MISTAKES OR MISCONCEPTIONS

HOW DOES THIS RELATE TO OTHER MATH TOPICS?

Math Action Terms/Words

Formulas	Equations

EXIT TICKET: THREE THINGS I LEARNED/WHAT STOOD OUT

1	
2	
3	

NEED PRACTICE	ALMOST GOT IT	GOOD AT THIS SKILL

TEACHER NOTE. PARENT NOTE

Made in United States
Orlando, FL
05 November 2026

90230219R00057